RAISING THE
HOME DUCK FLOCK

RAISING THE HOME DUCK FLOCK

A Complete Guide

Dave Holderread

Illustrated by Millie Holderread

 A Garden Way Publishing Book

STOREY COMMUNICATIONS, INC.
POWNAL, VERMONT 05261

 This book has been produced in cooperation with
The Hen House, PO Box 492, Corvallis, Oregon 97330.

Printed in the United States by Edwards Brothers

Fourth Printing, March, 1987

Library of Congress Cataloging in Publication Data

Holderread, Dave.
 Raising the home duck flock.

 First published in 1978 under title: The home duck flock.
 Bibliography: p.
 Includes index.
 1. Ducks I. Title.
SF505.H64 1979   636.5'97   80-10992
ISBN 0-88266-169-8

Dedicated to Mom —

the one who took me
— as a young child —
on walks through nature,
teaching me to smell the flowers
and listen to the song of birds. . .

Contents

TABLES

Foreword

Hobbyists, homesteaders and commercial duck raisers alike will find this book a handy reference. It is based on the author's longtime experience and supplemented with data from current scientific research. All aspects of small-scale duck husbandry are presented in an orderly, understandable form. Whether you are a beginner with a first-time pet, or a commercial operator wishing to formulate or mix your own feed, *Raising the Home Duck Flock* can provide the information you need.

AL HOLLISTER
Department of Poultry Science
Oregon State University
Corvallis, Oregon

Acknowledgments

This book was made possible because many people willingly shared their time, knowledge and resources. I want to take this opportunity to recognize those persons who played major roles in the preparation of *Raising the Home Duck Flock*. Special thanks to:

Wilbur Holderread, my father, for teaching me the basics of animal husbandry and giving helpful suggestions on the manuscript.

Frank Braman, lifelong poultry breeder, for patiently answering my questions over the years and for checking the accuracy of the manuscript.

Carol Glick, director of Academia Menonita Betania, for planting the idea and encouraging the writing of a manual for the home poultry flock owner.

The Alfredo Bonilla family of Aibonito Hatchery, for donations of equipment, birds and time to the vocational poultry program.

Stan and Fern Miller of Aibonito, Puerto Rico, for allowing Betania to convert their farm ponds into a waterfowl laboratory.

Larry and Cathy Passmore, for donating five months of their time to the Betania program.

Charles Lanpheare, retired coal miner and neighbor, for sharing old-time "secrets" on raising poultry.

Al Hollister, George Arscott and Dr. James Harper, of the Oregon State University Poultry Department, for supplying important references and offering useful suggestions on the manuscript.

Dr. Robert Jarvis, assistant professor of wildlife ecology, Fisheries and Wildlife Department, Oregon State University, for invaluable assistance on the section on sexing ducklings.

Mrs. John Heggen, Richard Heggen and Shirley Yoder for proofreading the manuscript and giving constructive criticism.

John C. Kriner, Jr., Orefield, Pa., outstanding waterfowl breeder, for information on the Australian Spotted breed.

*Henry K. Miller, Lebanon, Pa., American Poultry Association licensed

---

*Deceased

judge and prominent waterfowl breeder, for allowing us to photograph his birds and providing information on the Australian Spotted breed.

The many breeders and hatchery people who answered our inquiries and allowed us to photograph their birds.

Wanita Miller of Greenwood, Del., for working with us in the final months, helping with proofreading and typing.

And most of all, to Millie, my wife and best friend, for patience and encouragement throughout this project, for weighing feed and birds and keeping records on a number of experiments, for proofreading and typing the manuscript, and for the best part of the book—the sketches.

Introduction

In our world, approximately 25 percent of the human population suffer from serious nutritional deficiencies. Every day, including today, thousands die of starvation and diseases related to malnourishment. Many of us find these facts distressing as well as difficult to comprehend. And yet, they are a stark reality.

The problem of producing sufficient food is complex and many-faceted. There are no easy solutions. However, it has become obvious that modern agribusiness with its impressive arsenal of insecticides, herbicides, chemical fertilizers, monstrous machinery and large, specialized farms is not the ultimate answer we once thought it was.

The inability to produce and distribute adequate quantities of food is not the only shortcoming of current agricultural technology. In many cases production methods are depleting soil fertility and polluting our land, water and air at a dangerous rate, the quality of some of its products is questionable and vast numbers of small farmers are being strangled out of existence.

Growing numbers of people are becoming convinced that if we are to improve both the quantity and quality of food and at the same time restore our environment to a healthy condition, the emphasis must move back to smaller, less specialized farms. A practical result of this conviction is the increasing number of persons who are raising a major portion of their food.

Due to their size, efficiency and adaptability, poultry are the most popular type of small stock raised by homesteaders and backyard farmers. In many countries, chickens are the most common fowl kept. However, after raising and studying the various types of domestic poultry, I believe that for many situations, ducks are the most practical bird available. In a world of limited space and resources, the practicality of raising ducks becomes obvious as one recognizes the unique abilities of these waterfowl to utilize foodstuffs that normally go unharvested, control harmful insects, slugs, snails and unwanted aquatic plants, thrive under harsh conditions with limited shelter, resist diseases and parasites, and produce food efficiently.

While several good publications on waterfowl are available for the exhibitor and commercial producer, literature prepared specifically for the home flock owner is limited. *Raising the Home Duck Flock* is an attempt to provide a comprehensive handbook that clearly outlines the information needed by backyard farmers to raise ducks efficiently and with the least number of frustrations.

Because my experience with ducks has been limited to tropical Puerto Rico and the Pacific Coast of the United States, I lack firsthand knowledge in caring for waterfowl in other regions. To help bridge this gap, veteran poultry breeders and hatchery people from across North America were corresponded with or interviewed in laying the groundwork for this book.

In various sections of the text, the material presented was obtained from work conducted through the vocational poultry program at Academia Menonita Betania in Aibonito, Puerto Rico. The studies carried out at Betania were specifically designed for gathering information which would be pertinent to the small, home poultry flock.

Although this manual emphasizes the practical aspects of ducks, I hope you will also learn to appreciate the beauty and entertainment they have to offer. Over the past twenty years, ducks have added a great deal of color to my daily life, as well as having provided food and clothing. I trust they'll do the same for you.

DAVE HOLDERREAD

CHAPTER 1

Why Ducks?

The popularity of ducks is increasing in many areas of the world. It appears that the rest of us are beginning to understand what Europeans and Asians have known for centuries—ducks are one of the most *versatile* and *useful* of all domestic animals.

For the home poultry flock, we're looking for birds that produce meat and eggs efficiently, require minimal care and shelter, can find a good portion of their own food, destroy weed seeds and insects, don't need to be vaccinated and otherwise pampered, and add a touch of beauty and interest to our lives. For many situations, it's difficult to find a better all-purpose bird than *the duck*.

EASY TO RAISE

People who have kept all types of poultry generally agree that ducks are the easiest domestic bird to raise. Along with guinea fowl and geese, ducks are incredibly resistant to disease. While chickens usually must be vaccinated for communicable diseases and regularly treated for worms, coccidiosis, mites

1

and lice, duck keepers can normally forget about these inconveniences. Even when kept under poor conditions, small duck flocks are seldom bothered by sickness or parasites.

TABLE 1
GENERAL COMPARISON OF POULTRY

Bird	Raisability	Disease Resistance	Special Adaptations
Coturnix Quail	Good	Good	Egg and meat production in extremely limited space.
Guinea Fowl	Fair-Good	Excellent	Gamy-flavored meat; insect control; alarm. Thrive in hot climates.
Pigeons	Good	Good	Message carriers; meat production in limited space. Quiet.
Chickens	Fair-Good	Fair-Good	Eggs; meat; natural mothers. Adapt to cages, houses or range.
Turkeys	Poor-Fair	Poor-Fair	Heavy meat production.
Ducks	Excellent	Excellent	Eggs; meat; feathers; insect, snail, slug, aquatic plant control. Cold, wet climates.
Geese	Excellent	Excellent	Meat; feathers; lawn mowers; "watchdogs"; aquatic plant control. Cold, wet climates.

RESISTANT TO COLD, WET AND HOT WEATHER

Mature waterfowl are practically immune to wet or cold weather and better adapted to cope with these conditions than are chickens, turkeys, guineas or quail. With thick coats of well-oiled feathers, ducks can remain outside in the wettest weather. While chickens have protruding combs and wattles that must be protected from frostbite, and bare faces that allow the escape of valuable body heat, ducks are more completely clothed and are able to winter comfortably outdoors with only a windbreak, even when temperatures fall below 0°F. Ducks also do well in hot climates if they have access to plenty of shade and drinking water. During torrid weather, bathing water is beneficial.

INSECT, SNAIL AND SLUG EXTERMINATORS

Nurturing a special fondness for mosquito pupae, Japanese beetle larvae, potato beetles, grasshoppers, snails and slugs, ducks are extremely effective in controlling these and other pests. In the midwestern United States, ducks are used to reduce plant and crop damage during severe grasshopper infestations. In areas plagued by liver flukes, ducks can help correct the problem by consuming the snails which host this troublesome livestock parasite.

Under most conditions, two to six ducks per acre of land are needed to get rid of Japanese beetles, grasshoppers, snails and slugs. To free bodies of water of mosquito pupae and larvae, six to ten ducks should be provided for each acre of water surface. The breeds of ducks in the egg and bantam categories are the most active foragers, making them the best exterminators.

PRODUCTIVE

Ducks are one of the most efficient producers of animal protein. Indian Runner and Campbell duck hens selected for egg production lay *better* than the best egg strains of chickens, averaging 275 to 325 eggs per hen per year. Furthermore, duck eggs are 20 to 35 percent larger than chicken eggs. Unfortunately, in many localities, strains of ducks that have been selected for top egg production are not as readily available as egg-bred chickens. Meat-type ducks that are raised in confinement are capable of converting 2.6 pounds of concentrated feed into one pound of bird. If allowed to forage where there is a good supply of natural foods, they can do considerably better. The only domestic animal commonly used for food that has better feed conversion is the chicken broiler with a 2.1:1 ratio. See Tables 2 and 3 for a comparison of the producing abilities of all types of poultry.

TABLE 2
COMPARATIVE EGG PRODUCTION OF POULTRY

Bird	Egg Weight per Dozen ounces	Annual Egg Production #	pounds	Annual* Feed Consumption pounds a	b	Feed to Produce 1 Pound of Egg pounds	Efficient Production Life of Hens years
Duck, Campbell	31.0	288	46.5	110	140	2.4-3.0	2-3
Quail, Coturnix	5.5	306	8.8	—	22	2.5	1
Chicken, Leghorn	24.0	240	30.0	85	95	2.8-3.2	1-2
Duck, Pekin	40.0	156	32.5	145	175	4.5-5.4	2-3
Goose, China	66.0	72	24.8	130	185	5.2-7.5	4-8
Guinea Fowl	17.0	78	6.9	45	65	6.5-8.7	1-2
Chicken, Broiler	25.0	144	18.8	125	130	6.6-6.9	1-2
Turkey, Lg. White	38.0	90	17.8	180	200	10.1-11.2	1-2

Based on the egg yields of good stock fed concentrated feeds and managed for efficient production. Egg size, egg production and feed conversion of poultry are highly dependent upon the quality of the birds and the care they receive.

*Represents the quantity of feed that the average hen consumes from the day she commences laying until a year later.

a—For hens that are free to roam and forage.

b—For hens raised in confinement.

TABLE 3
COMPARATIVE MEAT PRODUCTION OF POULTRY

Bird	Optimum Butchering Age weeks	Average Live Wt. at Butchering pounds	Feed Consumption pounds	Feed to Produce 1 lb. of Bird pounds
Chicken, Broiler	8	4.0	8.5	2.1
Duck, Pekin	7	7.0	19.0	2.7
Turkey, Lg. White	16-20	17.0	55.0	3.2
Goose, China	12-20	10.0	35.0	3.5
Quail, Coturnix	6	.4	1.5	3.8
Guinea Fowl	12-18	2.3	11.0	4.8
Duck, Campbell	10-14	4.0	22.0	5.5
Chicken, Leghorn	18-22	3.3	20.0	6.1

Based on meat yields of quality stock which were fed concentrated feeds, kept in confinement and managed for efficient production. Growth rate and feed conversion are highly dependent on the quality of the birds and their care.

EXCELLENT FORAGERS

Ducks are energetic foragers, capable of rustling 15 to 100 percent of their own living. Along with guineas and geese, ducks are the most efficient type of domestic poultry for the conversion of food resources that normally are wasted, such as insects, weed plants and seeds, into edible human fare.

AQUATIC PLANT CONTROL

Ducks are useful in controlling unwanted plants in ponds, lakes, and streams, improving conditions for many types of fish. In most situations, twelve to twenty birds per acre of water are required to clean out heavy growths of green algae, duckweed (Lemna), pondweed (Potamogeton), widgeon grass (Ruppia), muskgrass (Chara), arrowhead (Sagittaria), wild celery (Vallisneria), and other plants that ducks consume. Once the plants are under control, four to ten ducks per acre will usually keep the vegetation from taking over again.

In bodies of water having plants submerged more than two feet, or when it is desirable to clean grass from banks, four to eight geese per acre of water surface should be used along with the ducks. (For geese to be effective, they must be confined to the pond and its banks with fencing.) Waterfowl are *ineffective* in checking the growth of tropical plants such as water lettuce and water hyacinth.

GARBAGE DISPOSAL

Ducks will eat almost anything that comes out of the kitchen or root cellar. They relish vegetable trimmings, table and garden leftovers, canning and fruit juice refuse, most kinds of deteriorated produce, and stale baked goods. To make it easier for these broad-billed fowl to eat firm vegetables and fruits, place apples, potatoes, beets, turnips, etc. on an old board and crush them with your foot.

In the poultry program in Puerto Rico, we supplied a flock of forty Rouen ducks that were on pasture and had access to a five-acre pond, with nothing but leftovers from the school cafeteria. These garbage-fed birds remained in good flesh and showed no signs of poor health, although they produced 60 percent fewer eggs than a control group that was provided concentrated laying feed along with limited quantities of institutional victuals.

USEFUL FEATHERS

The down and small body feathers of ducks are valuable as filler for pillows and as lining for comforters and winter clothing. (See Appendix F, page 157.)

VALUABLE MANURE

A valuable by-product of raising ducks is the manure they produce. Duck manure is an excellent organic fertilizer that is high in nitrogen. The innovative Orientals have taught us another use for this "waste." They herd their duck flocks through rice fields to eat insects and pick up stray kernels of grain. The birds are then put on ponds where their manure provides food for fish.

GENTLE DISPOSITIONS

Rather shy, ducks are seldom aggressive towards humans. Of the larger domestic birds, they are the *least* likely to inflict injury to children or adults. In the twenty years I have worked with ducks, the only injuries I've sustained have been small blood blisters on my arms and hands—received while attempting to remove eggs from under broody hens—and an occasional scratch when the foot of a held bird escaped my grasp. If you do get scratched by the claws of a bird, prompt washing of the wound with rubbing alcohol will lessen the chance of infection.

DECORATIVE AND ENTERTAINING

Together with their many practical attributes, ducks are decorative and entertaining. A small flock of waterfowl can transform just another pond into a center of attraction and provide hours of entertainment.

TABLE 4
COMPARISON OF DUCKS AND CHICKENS

Characteristic	Ducks	Chickens
Housing requirements	Minimal	Substantial
Height of fence to keep confined	2-3 feet	4-6 feet
Susceptibility to predators	High	Moderately high
Resistance to parasites and disease	Excellent	Fair
Resistance to hot weather	Good	Good
Resistance to cold weather	Good	Fair
Resistance to wet weather	Excellent	Poor
Foraging ability	Good-Excellent	Fair-Good
Scratching in dirt	None	Considerable
Probing in mud with bills or beaks	Considerable	None
Incubation period	28 days	21 days
Hatchability of eggs in incubators	65-90 percent	70-90 percent
Cost of day-olds in lots of 25	Ducklings double the price of chicks	
Starter feed—min. protein needed	16 percent	18 percent
Layer feed—min. protein needed	15 percent	15 percent
Age hens commence laying	20-24 weeks	20-24 weeks
Eggs laid per hen per year (wt.)	32-52 lbs.	22-34 lbs.
Feed to produce 1 pound of eggs	2.4-3.8 lbs.	2.8-4.0 lbs.
Part of diet hens can forage	10-25 percent	5-15 percent
Light required for top production	14-16 hours	14-17 hours
Annual mortality rate of hens	0-3 percent	5-25 percent
Efficient production life of hens	2-3 years	1-2 years
Protein content of eggs	13.3 percent	12.9 percent
Fat content of eggs	14.5 percent	11.5-12.5 percent
Cholesterol content of eggs	.884 percent	.494-.638 percent
Flavor of eggs	Similar; ducks sometimes stronger	
Feed to produce 1 pound of bird	2.5-3 lbs.	2-2.2 lbs.
Age to butchering	7-12 weeks	8-20 weeks
Typical plucking time	3-15 minutes	2-10 minutes
Color of meat	All dark	Light and dark
Protein content of flesh	21.4 percent	19.3 percent
Fat content of carcass	16-30 percent	5-25 percent
Cholesterol content of carcass	0.07 percent	0.06-0.09 percent
Usefulness of feathers and down	Excellent	Fair

A domestic Mallard hen with ten-day-old ducklings.

I can still remember the first ducks I had as a young boy. Because our property had no natural body of water, I fashioned a small dirt pond in the center of the duck yard. After filling it with water, I watched as my two prized ducklings jumped in and indulged in their first swim. They played and splashed with such enthusiasm that it wasn't long before I was as wet as they were. And then—much to my delight—they began diving, with long seconds elapsing before they popped above the surface in an unexpected place. I was "hooked," and continue to be intrigued by the playfulness, beauty and grace of swimming waterfowl.

CHAPTER 2

Some Points to Consider

Despite the versatility and usefulness of ducks, there are several factors you should be aware of and understand. In some situations another type of fowl may prove more suitable.

NOISE

Many people find the quacking of ducks an acceptable if not pleasant part of nature's choir. However, if you have close neighbors, the gabble of talkative hens may not be appreciated or tolerated. Some breeds are noisier than others. A small flock consisting of one of the calmer breeds will be quiet if not frightened or disturbed frequently. Drakes are almost mute, and for the control of slugs, snails and insects in town or suburb, they work fine.

PLUCKING

Most people find that plucking a duck is more time-consuming than defeathering a chicken. But then, most of us who have had the privilege of dining on roast duck agree that it is time well spent. Furthermore, duck feathers are much more useful than chicken plumes. With good technique and a little experience, it is possible to reduce the picking time to three minutes or less.

POND DENSITY

Having large numbers of ducks on small ponds or creeks encourages unhealthy conditions and can result in considerable damage to bodies of water. One of the feeding habits of ducks is to probe the mud around the water's edge for worms, roots and other buried treasures. A high density of ducks will muddy the water and hasten bank erosion. On the other hand, a reasonable number of birds (15 to 25 per acre of water) will improve conditions for fish, control aquatic plant growth and mosquitos, and will not significantly increase bank erosion.

GARDENS

Putting a few ducks (the bantam breeds are best due to their small size) in a garden is a good method of controlling slugs, snails and insects. However, to prevent the birds from doing more harm than good, the following guidelines must be observed:

1. Don't let birds in until the crops are well started and past the succulent stage.
2. Keep ducks out when irrigating or when the soil is wet.
3. Fence off tender crops such as lettuce, spinach, cabbage and green beans.
4. Remove birds when low-growing berries and fruits are ripe.
5. Limit the number of ducks in a medium-sized garden to two or three adults.

MEAT AND EGGS

All types of poultry provide good food. However, there are variations in the flavor, texture and composition of the meat and eggs produced by the diverse species. There are also differences in dietary needs and likes and dislikes of food among people. If you are seriously considering producing duck meat or eggs but have never cooked or eaten them, I recommend that you sample duck products before starting your flock. (This practice is wise before investing time and resources in any type of unfamiliar animal for food.) The following observations are presented to help you evaluate your first encounter with duck cuisine:

1. The flavor of meat and eggs from ducks whose diet included fish or fish products is often strong and *not* typical of good duck.
2. Ducklings which have been raised in close confinement and pushed for top growth (e.g., Long Island Ducklings) are much fatter than ducks which have foraged for some of their own food and have grown at a

slower pace. The high fat content of quick-grown duckling makes its meat exceptionally succulent and provides valuable energy for persons who live in cold climates or who get strenuous physical exercise. However, many of us get limited exercise and our bodies do not need the large quantity of fuel provided by fat ducklings. For this reason, I feel

TABLE 5
APPROXIMATE COMPOSITION OF VARIOUS MEATS

Description	Calories per 100 grams	Protein percent	Fat percent
Chicken: raw			
Fryers			
Total edible	155	17.3	7.4
Flesh only	107	19.3	2.7
Roasters			
Total edible	239	19.2	17.9
Flesh only	131	21.1	4.5
Mature hens and cocks			
Total edible	298	17.4	24.8
Flesh only	155	21.6	7.0
Turkey: raw			
Total edible	218	20.1	14.7
Flesh only	162	24.0	6.6
Duckling: domestic, raw			
Total edible	326	16.0	28.6
Flesh only	165	21.4	8.2
Duck: wild, raw			
Total edible	233	21.1	15.8
Flesh only	138	21.3	5.2
Rabbit: raw			
Flesh only	162	21.0	8.0
Pork: raw			
Carcass (medium-fat class)			
Total edible	513	10.2	52.0
Composition of trimmed lean cuts, ham, loin, shoulder and spareribs (medium-fat class)			
Total edible	308	15.7	26.7
Beef: raw			
Carcass			
Total edible			
Choice grade	379	14.9	35.0
Good grade	323	16.5	28.0
Total edible, trimmed to retail levels			
Choice grade	301	17.4	25.1
Good grade	263	18.5	20.4
Round			
Total edible	197	19.5	12.5

Information from *Handbook of the Nutritional Contents of Foods.*

TABLE 6
APPROXIMATE COMPOSITION OF EGGS

Kind of Egg	Protein percent	Fat percent	Cholesterol* mg/gm of egg	Calories per 100 gms
Chicken, Commercial Egg	12.9	11.5	4.94-5.50	163
Chicken, Commercial Broiler	—	—	6.38	—
Quail, Coturnix	11.5	10.9	8.44	186
Turkey	13.1	11.8	9.33	170
Duck	13.3	14.5	8.84	191
Goose	13.9	13.3	—	185

Information from *Handbook of the Nutritional Contents of Foods* and "Cholesterol Content of Market Eggs," *Poultry Science Journal.*
*Young hens produce eggs with less cholesterol than old hens.
Note: Dashes indicate that information was not available.

that quick-grown duckling should be eaten only in limited quantities. See page 82 for ways to produce lean ducks.

3. Duck eggs have a higher cholesterol content than the average chicken egg. As with fat in meat, the amount of cholesterol in eggs seems to be affected by the diet and life-style of the producing bird. Hens that are active and forage for a portion of their diet *may* produce eggs lower in cholesterol.

    When eggs are eaten in moderation (we limit our weekly intake to four to six eggs per person), the difference in cholesterol between duck and chicken eggs probably is insignificant for healthy people who get adequate exercise and eat sensibly—lots of high fiber and uncooked foods while going easy on the meats.

4. Duck eggs are excellent for general eating and baking purposes. When fried in an open pan, the whites of duck eggs are often firmer than those of chicken eggs. We prefer to steam-fry our duck eggs (see Appendix E, page 153) for a softer textured egg white. While it is often said that duck eggs are unsatisfactory for meringues and angel food cakes, we have not found this statement to be true. In fact, one of our favorite treats for special occasions is whole wheat angel food cake made with duck egg-whites. (See Appendix E.)

CHAPTER 3

External Features

Ducks are masterfully designed, down to the smallest detail, for an aquatic life. Being acquainted with the external features of ducks is a useful management tool and will deepen your appreciation for these waterfowl. I suggest that you take a few minutes to familiarize yourself with the accompanying nomenclature diagram.

BODY SHAPE

For stability and minimal drag while swimming and diving, the lines of the body are smooth and streamlined; the underbody is wide and flat.

FEATHERS

Endowed with an extraordinary abundance of feathers, particularly on the underside of their bodies, ducks can swim in the coldest water and remain comfortable. Soft, insulating down feathers close to the skin are covered by larger contour feathers. Several times each day, ducks faithfully preen and oil their feathers. A water repellent is produced by a hidden oil gland at the base of the tail. As a duck preens, it frequently squeezes the nipple of the oil gland with its bill and spreads the excretion onto its feathers.

WINGS

Ducks have long, pointed, rather narrow wings—except Muscovies, whose wings are broad and rounded. Most domestic breeds have lost their ability to fly, although Muscovies, Calls, East Indies and domesticated Mallards have retained their flying skills to varying degrees.

TAILS

Muscovies have squarish tails that are 4 to 6 inches long. Other ducks possess shorter, pointed rudders. Excluding the Muscovy, mature drakes have several curled feathers in the center of their tails.

BILL

Long and broad, duck bills are well adapted for collecting food from water, catching flying insects and rooting out underground morsels. With nostrils located near the head, ducks can dabble in shallow water and breathe at the same time. Due to hormonal changes in their bodies, duck hens with yellow or orange bills frequently develop dark spots or streaks on their bills when they begin to lay. These blemishes are not a sign of disease. Some waterfowl breeders believe they are an indication of high egg production.

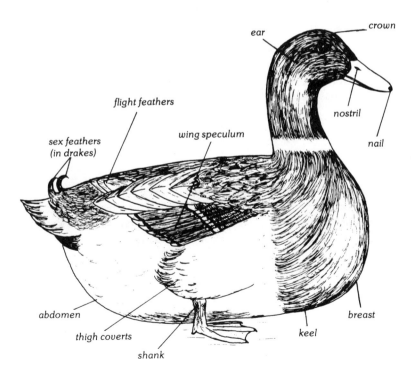

With their long, broad bills, ducks are adept at catching flying insects, straining minute plants and animals from water and probing in mud. The hooked nail at the end of the bill is used for nipping off tough roots and stems.

EYES

The vision of ducks is much sharper than our own. Due to the location of their eyes, ducks can see nearly 360 degrees without moving their heads. This feature makes it possible for a feeding duck to keep a constant lookout for danger.

FEET AND LEGS

With webbed toes and short legs, ducks are not exceptionally fast on land, but are highly skilled swimmers, preferring to escape danger by way of water. The legs of ducks are injured more readily than those of chickens. Therefore it is advisable to pick them up by the neck and/or body.

BACK

Normally a duck's back is long and straight. In some of the larger breeds— particularly Aylesburies and Rouens—there is a tendency toward arched backs. These humped backs can be a serious problem since ducks with this fault often have low fertility or are sterile. When selecting breeding stock, avoid individuals with excessively curved backs.

KEEL

Ducks that are heavily fed and selected for large size and fast growth often develop keels. This appendage is a fold of skin that hangs from the underbody, and in extreme cases may run the entire length of the body and brush the ground as the bird walks. A well-developed keel is a breed characteristic of Aylesburies and Rouens, while it is not preferred on any other breed.

CHAPTER 4

Behavior

Just as you and I have distinct personalities, so each duck has its own peculiarities and habits. However, ducks follow behavioral patterns which you should understand if you're going to do a good job of raising them.

PECKING ORDER

This bird law regulates the peaceful coexistence of the duck flock. The number one bird in the flock can peck or dominate all others, the number two bird can dominate all but number one, the number three bird can dominate all but numbers one and two, and on down the line until we reach the last individual who dominates no one.

If you introduce a new bird into an established flock, the existing pecking order is threatened and normally this results in a power struggle which may evoke fighting. Unless the conflict is causing serious injury to the participants, you should not intervene, remembering that the roughhousing is necessary for the future peace of the flock.

FEEDING

The natural diet of ducks consists of approximately 90 percent vegetable matter (including seeds, berries, fruits, nuts, bulbs, roots and succulent grasses)

16

and 10 percent animal matter (such as insects, mosquito larvae, snails, slugs, leeches, worms, and an occasional small fish or tadpole). Sand and gravel are picked up to serve as grinding stones in the gizzard. Ducks feed by dabbling and tipping up in shallow water, drilling in mud and foraging on land. While ducks eat considerable quantities of tender grass, they're not true grazers as are geese.

SWIMMING

Skillful and enthusiastic swimmers from the day they hatch, ducks will spend many happy hours each day bathing and frolicking on water if it is available. However, ducks of all types and ages (particularly Muscovies and ducklings) can drown if their feathers become soaked and they are unable to climb out of the water. For this reason, ducks must be allowed to swim only where they will be able to exit easily. It is a good idea to keep ducklings out of water until they are four weeks old. Ducks can be raised successfully without swimming water. Most of the ducks we've raised have had only drinking water.

MATING

Ducks will pair off, although domesticated drakes normally mate indiscrim-inately with hens in a flock. If you raise several breeds and wish to hatch pure-bred offspring, each variety needs to be penned separately three weeks prior to and throughout the breeding season.

In single male matings, a drake can usually be given two to five hens, al-though males sometimes have favorites and may not mate with the others. In flock matings, one medium or heavyweight breed male should be allowed for every three to six hens, while one drake of the lightweight breeds should be provided for every four to seven hens. For good fertility, hens need to be bred by drakes at least once every four or five days. A few eggs which have been laid as long as two weeks after copulation may hatch.

Ducks prefer to mate on water, but most breeds can copulate successfully on land. Several of the large breeds, especially deep-keeled Aylesburies and Rouens, have higher fertility if they have access to swimming water at least eight inches deep.

NESTING

Domestic duck hens frequently lay their eggs at random on the ground, and occasionally even while swimming. When a nest is made, it is a shallow de-pression in the ground, lined with twigs, grass and leaves. If eggs are left in

Mallards and their derivatives frequently tip up when feeding in water.

the nest for natural incubation, the hen will pluck down from her breast for added insulation.

FIGHTING

Ducks get along together well and rarely fight. If a new bird is introduced into a flock, there will be a short period of chasing, pushing and wing slapping, but normally such conflicts subside quickly. Adult ducks seldom inflict injury to one another if there is not an excessive number of drakes in a flock during the breeding season. As a rule, drakes will not fight among themselves if there are no females around.

LIFE EXPECTANCY

Ducks live a surprisingly long time when protected from accidental deaths. It is not unusual for ducks to live and reproduce for six to eight years, and there are reports of exceptional birds living twenty to thirty years. However, few ducks die of old age (except for those special pets!) since fertility decreases after one or two years and egg production normally declines significantly after three or four years, and their owners do not consider it economical to keep them past this age.

CHAPTER 5

Selecting a Breed

Choosing an appropriate breed for the home flock can be fun and will play an important role in the success or failure of your duck project. Unfortunately, novices often assume that a duck is a duck and acquire the first web-footed bird they find that quacks. This mistake frequently results in expensive eggs or meat, needless problems and a discouraged duck keeper. Investing a little time at the *outset* in acquainting yourself with the basic characteristics, attributes and weaknesses of the various breeds will go a long way toward eliminating unnecessary disappointments.

SOME CONSIDERATIONS

The following questions are designed to help you identify the features needed in birds for your flock. You may find it helpful to jot down your answers and then keep them handy as you compare the breeds.

Purpose

What is your main purpose for raising ducks? Is it for eggs, meat, feathers, insect and slug control or a combination of these and other aims?

Location

Where are you located? Some breeds are noisier than others—a fact which should be taken into consideration when you have close neighbors. Talkative breeds (Calls and Pekins in particular) also attract more predators.

Management

How are you going to manage the flock? Will they be confined to a small pen or be given freedom to roam over a large area? Will you provide all of their feed, or will they forage for a good portion of it themselves?

Color

What plumage color is best adapted to your situation? Aside from personal preferences, color is significant for several practical reasons. The pin feathers of light-plumaged birds are not as visible as those with dark plumage, making it easier to obtain an attractive carcass with light-colored ducks when they are butchered for meat. On the other hand, dark birds are better camouflaged, making them less susceptible to predators. Also, if there is no bathing water available, colored ducks maintain a neater appearance than white ones.

Availability

What breeds are available either locally or by mail order? Some breeds are rare, making them more expensive and difficult to obtain.

THE BREEDS

Over the years, waterfowl breeders in various parts of the world have developed distinctive types of ducks. As these local ducks became uniform in size, shape and color, they were given names and recognized as separate breeds. Those breeds which have proven to be the most useful and widely accepted have survived.

Today, most of the purebred ducks being raised in North America belong to one of sixteen breeds. Many of these breeds consist of several varieties that are distinguished from one another by the color or pattern of their plumage. For convenience, the breeds are divided into four main categories: egg, meat, general purpose and bantam.

Egg Breeds

Ducks that are bred primarily for high egg production are small, exceptionally active, non-broody, and often are more high-strung than other breeds. Despite their diminutive size, they lay a greater number of eggs than larger

TABLE 7
BREED PROFILES

Category	Breed	Weight in Pounds M	F	Yearly Egg Production	Egg Size per Dozen Ounces	Mothering Ability	Foraging Ability	Availability
Egg	Bali	5.0	4.5	150-250*	30-36	Poor	Good	Poor
	Campbell	4.5	4.5	250-325+	28-34	Poor	Excellent	Good
	Runner	4.5	4.0	225-325+	30-36	Poor	Excellent	Good
Meat	Aylesbury	9.0	8.0	35-125	36-42	Fair	Fair	Fair
	Muscovy	12.0	7.0	50-125	44-50	Excellent	Excellent	Good
	Pekin	9.0	8.0	125-175	38-42	Poor	Fair-Good	Excellent
	Rouen	9.0	8.0	35-150	36-44	Fair	Fair-Good	Excellent
General Purpose	Cayuga	8.0	7.0	100-175	34-38	Good	Good	Fair
	Crested	7.0	6.0	100-175	34-38	Fair	Good	Fair
	Magpie	6.0	5.0	125-225	30-36	Poor	Good	Poor
	Orpington	8.0	7.0	150-250	30-36	Poor	Good	Fair
	Swedish	8.0	7.0	100-150	34-40	Good	Good	Fair
Bantam	Australian Spotted	2.5	2.2	25-100*	22-30	Excellent	Excellent	Poor
	Call	1.8	1.6	25-75	18-24	Excellent	Excellent	Good
	East Indie	2.0	1.8	25-125	18-28	Excellent	Excellent	Fair
	Mallard	2.8	2.4	25-125	24-30	Excellent	Excellent	Excellent

Information presented in this profile is based on the average characteristics of each breed. Actual performance of individuals may vary considerably from the norm.

*Estimates

hens, while consuming a lot less feed. Excellent foragers, they are capable of picking up much of their own food, and are outstanding for the control of insects, snails and slugs. While all ducks enjoy swimming water, the egg breeds are more at home on land than most. If *efficient egg production* is your principal goal, one of these breeds should be given top priority.

Bali. In my opinion, the Bali is the most unusual of all breeds. With their dignified, yet comical appearance, they can bring a smile to the face of even the most serious-minded person. One of the oldest of domesticated ducks, they originated on Bali and other Indonesian islands. Waterfowl historians generally agree that this breed and Indian Runners are closely related; the Runner was probably developed from the Bali.

The erect carriage of the Bali is accentuated by a long body that is somewhat heavier than that of the Indian Runner. Resting snugly on the head of an ideal Balinese is a medium-sized globular crest. As with Crested ducks, the ducklings of this breed hatch out both plain-headed and crested. Most Bali raised in North America are white, although khaki brown is the common color in their homeland.

While Bali are good layers, they are so scarce outside their homeland that they have not been selectively bred for egg production. However, if you can locate specimens, it is a practical and fascinating breed with which to work.

A White Bali drake. Notice his thicker neck and heavier-set body when compared with the Indian Runners on page 25.

Campbells. Mrs. Adele Campbell of Glouchester, England, is given credit for having developed the colorful Khaki Campbell in about 1900. It is believed that Indian Runners, Mallards and Rouens were used in its creation. Egg-bred Campbells are the best layers of all ducks. There have been reports of individuals producing 360 or more eggs in a calendar year, although flock averages of 275 to 340 are more typical. Campbell eggs have superb texture and flavor and pearly white shells.

Here is an appropriate place for a word of caution: If you want good egg yields, make certain you acquire authentic Campbells that have been selected for high egg production. The laying ability of birds in some hatchery flocks has been allowed to deteriorate, and much too often crossbreeds are being sold as Campbells.

Today, there are three varieties of Campbells—White, Dark and Khaki. While the Khaki is generally considered by far the most common, a substantial number of so-called Khakis more accurately are Darks. Khaki Campbell drakes have green or bluish-green bills, greenish-bronze heads and backs (brownish-bronze in these sections is preferred for exhibition), with the remainder of the plumage being an even shade of khaki tan that is similar in color to withered grass. Ducks have dark brown or greenish-black bills, and except for seal brown heads, the plumage should be a warm shade of khaki brown. When compared to Khakis, Dark Campbell females display deeper

Pair of egg-bred Khaki Campbells exhibiting the good length of body that is needed for top egg production. Owned by Mother Hen Farm, Corvallis, Oregon.

White Campbell females. This variety can be distinguished from other white ducks by their trim heads, medium size, and spry carriage, which is roughly 35 degrees above horizontal. Owned by Wilbur and Dave Holderread, Corvallis, Oregon.

shades of brown while males have brighter greenish-black heads and backs. Due to a sex-linkage factor, when Khaki drakes are mated to Dark ducks, their offspring are color-coded and can be sexed at hatching time by the hue of their down—males are dark, females are light.

Until recently, White Campbells were rarely heard of in North America. During the last several years, however, a number of waterfowl breeders have been distributing this useful variety nationwide. In our experience, Whites are as productive and hardy as their colored counterparts. Plumage color is silky white and feet are orange. The bill color of egg-bred White Campbells normally is pink in ducklings and pink or yellowish-pink in adults.

Campbells are extremely hardy and active, and will forage over a large area. Of all breeds of true ducks, they seem to have the least desire to swim. Campbells are also one of the quieter breeds, although this characteristic does vary among different strains. When dressed for meat, they provide a medium-small carcass that is lower in fat than most large ducks. The colored varieties have excellent camouflage and are pretty, especially when seen against the backdrop of a green pasture. For the home flock, I consider Campbells to be the best duck—and in many situations, the best bird—for the efficient production of eating eggs.

Indian Runners. Because of their long, slender bodies and vertical posture, this breed is often called the *Penguin duck.* Unlike most ducks, Indian Runners do not waddle—they run. Aided by their fast gait and active disposition, Runners are one of, if not the best foragers of all domestic breeds.

Many colorful varieties have been developed in this popular breed, including White, Penciled, Fawn and White, Solid Fawn, Blue, Black, Chocolate,

Indian Runners showing upright carriage, slender bodies, wedge-shaped heads and high-set eyes typical of the breed. A pair of Fawn and Whites are in the background, Whites in foreground. Owned by Henry K. Miller, Lebanon, Pennsylvania.

Buff, Khaki and Gray (same markings as the Mallard). The first three varieties are the most common.

The White, and Fawn and White Runners are the best layers since they have been selected more carefully for egg production than the other varieties. Runner eggs are white, tinted, green or blue, and average two ounces larger per dozen than those of Campbells.

Although Indian Runners are not as large as breeds traditionally used for meat, they are popular with people who desire a small, lean meat-bird. Second only to Campbells in egg production, Runners are a good breed for the home duck flock, being practical as well as highly entertaining. In fact, watching the astonished expressions on people's faces the first time they see Indian Runners stroll by is a pretty good reason for keeping a few.

Meat Breeds

The meat breeds can be identified by their large sizes, calm dispositions and fast growth rates. As a group, they are easy to keep confined, do not wander far from the feeding area and, with the exception of the Pekin, are less vocal than most smaller ducks. These birds are ideally suited for the efficient production of large roasting ducks. Due to their hearty appetites and only fair laying abilities, it is not practical to keep them solely for the production of eating eggs.

Aylesbury. This massive pure white duck is a native of England. The body carriage of Aylesburies is horizontal and their well-developed keels nearly touch the ground. They possess unusually long, straight bills which are pinkish-white. Their skin is white, in contrast to the yellow skin of most other breeds. Hens lay white, tinted or green eggs.

Aylesburies are slow moving, *exceptionally tame* and make large, fat roasting ducks when well-fed on concentrated feed. For top fertility and hatchability, breeders must have swimming water and should be fed a ration that contains 18 to 20 percent protein and is well fortified with vitamin-rich substances such as alfalfa meal and cod-liver oil or a vitamin premix. Generally, Aylesburies are not quite as hardy and vigorous as most other breeds.

Muscovy. A native South American, this duck is definitely a bird of a different feather. In fact, it is not considered a true duck by many waterfowl breeders, and has been grouped with them for lack of a better classification. Muscovies will cross with other ducks, but the resulting offspring are usually sterile.

Muscovies have many distinguishing features. Their faces are covered with rough, red skin, and drakes have a fleshy knob at the base of their bills. Both sexes display narrow head crests which can be elevated and lowered at will.

A pair of deep-keeled Aylesburies, popular in England where white-skinned fowl are preferred for meat. Owned by Henry K. Miller.

Hens are practically mute and the voices of males resemble a muffled hiss. Drakes are nearly twice as heavy as their mates and have larger facial skin patches, making it easy to distinguish the gender of mature birds.

While Muscovies do swim, their feathers are not as water repellent as those of genuine ducks. When forced to remain on water for more than a brief length of time, they can become soaked and drown. They require more protection from wet and cold weather than other ducks.

Muscovies often roost on buildings and walls. The hens are strong fliers, but drakes frequently become so ponderous that they cannot get airborne without the aid of a strong headwind or an elevated perch.

Muscovies are excellent foragers and can utilize larger quantities of grass than true ducks. In mild weather, ducklings that are brooded artificially can be put out on pasture at three to six weeks of age. If good quality forage is available, pastured ducklings require only light feedings of grain or pellets once daily. When raised in this manner, they'll average five pounds for hens and ten pounds for drakes at sixteen to twenty weeks of age.

The jumbo white or tinted eggs laid by the Muscovy require thirty-three to thirty-five days to hatch. Hens are dedicated natural mothers—with the exception of some production-bred white strains—and normally bring off large broods. It is not unusual to see Muscovy hens with fifteen to twenty ducklings in tow. They bravely protect their young, and have been known to ward off marauding predators such as foxes, cats and dogs.

A Black and White Pied Muscovy drake with his head crest elevated.

A White-headed Muscovy duck and Colored drake displaying his white forewings which are characteristic of Colored varieties. Owned by Mrs. Richard Moon, Bend, Oregon.

Muscovies are highly prized for their meat and are easier to pick than true ducks. They have exceptionally broad, well-muscled breasts and are one of the leanest of all waterfowl. We have found that when their meat is used in stews it is hard to distinguish from beef, while cured and smoked Muscovy is similar to lean ham.

The two most common varieties of Muscovies are the White and Colored, with the latter's plumage being iridescent greenish-black throughout, except for white forewings. Other varieties include Blue, Chocolate, Silver, Buff and Pied.

This breed is not considered attractive by many people, but they are hardy and quiet, and capable of foraging for much of their keep. If you want a meat-bird that requires minimal care and feed, it is doubtful that there's a better choice available than the Muscovy. An added bonus our family has found with this breed is that we don't become overly attached to them—making it less of an emotional strain on butchering day.

Pekin. Recognized worldwide as the fastest growing of all ducks, Pekins are hatched in greater numbers than all other meat type breeds combined. These hardy ducks were introduced into Europe and the Americas from China in the 1870s. Pekins are attractive as well as practical, and frequently

A group of seven-week-old Pekins. Note typical breed characteristics of large, high-crowned heads, rather upright carriage and elevated tails.

are kept on ponds "just for pretty." Their creamy-white plumage is offset nice-ly by bright orange bills and reddish-orange feet. Hens lay white or tinted eggs.

The Pekin is the best breed for the production of "green" ducklings. When managed properly, they are capable of weighing seven pounds at seven weeks of age. One objection some people have to this breed is the excep-tionally high fat content of their carcasses and the talkativeness of the hens.

Rouen. France was the original home of the imposing Rouen. Their large, deep bodies and handsome plumage have helped them gain wide popularity. They have retained the basic color pattern of the Mallard, although the plumage of the Rouen typically is several shades darker than their wild ances-tors. Hens lay eggs in various shades of blue, green and tinted white.

Rouens combine beauty with good meat qualities. They do not grow quite as rapidly as Pekins, making them less practical for raising as "green" duck-lings. Rouens are better suited for situations where they can forage for a sub-stantial portion of their food, receiving just enough grain or concentrated feed to keep them in good condition. They can be dressed at ten to twelve weeks or five to six months when they are well-matured and in full feather.

Deep-keeled standard-bred Rouens photographed in midsummer after the drake had commenced the eclipse molt. Owned by Marcus L. Davidson, Bath, Penn-sylvania.

A pair of production-bred Rouens. The drake has just begun the eclipse molt, as evidenced by the sprinkling of dark feathers on his gray underside.

General-purpose Breeds

This category includes a colorful assortment of medium weight breeds that are capable of furnishing a good supply of both meat and eggs for your kitchen. Not being the outstanding layers that egg-type ducks are, nor as fast growing as Pekins, these breeds are the most practical where they can rustle a good portion of their food. The fastest growing strains make good prospects for the production of ten- to twelve-week-old roasting ducklings. These breeds are active foragers and will add color and interest to your duck flock.

Cayuga. One of three breeds of ducks that can claim North America as its native land, the Cayuga is thought to have been developed from ducks raised in the region of Lake Cayuga, New York.

The distinguishing feature of this attractive bird is its brilliant green-black plumage. A common problem in breeding Cayugas is that the brightest colored individuals (particularly hens) frequently acquire white feathers as they age. To keep this problem under control, the breeder of Cayugas can select good green-colored drakes with no white feathers (be sure to check under the bill) and mate them with hens that had no white feathers as young birds. Hens with just a hint of brown under their bills and wings often produce ducklings with solid black color.

In keeping with her ebony plumage, the hen produces her first eggs of the season encased in a black or dark gray film that fades to light gray or blue as the season progresses. Ducklings have glossy black feet and bills, and black down with just a trace of yellow on the breast.

An old Cayuga duck with good solid black plumage. Owned by Maurice Clark, Toledo, Oregon.

Cayugas are one of the hardiest and quietest of all breeds. However, their black plumage makes it difficult to obtain an attractive carcass when they are butchered at a stage when they have pin feathers. (When pin feathers are present, we solve the problem by skinning the bird. This procedure is more healthful, since most of us don't need the skin's extra fat in our diet.)

Crested. A large, round head crest gives this regal bird its name. Over the years that domestic ducks have been raised, specimens with small tufts have frequently appeared in many countries. These tufted ducks were improved through selective breeding in various parts of the world, including Europe and North America, resulting in the Crested duck as we know it today.

A curious trait of Crested ducks is that the size and shape of the crest still varies a great deal, and up to one-third of all ducklings lack the head adornment. This phenomenon is the result of a lethal gene that causes the premature death of embryos that carry two genes for crest. The only ducklings that can hatch are those which have only one gene for crest or else have two genes for normal (no crest). Due to this condition, the average hatchability of Crested duck eggs is approximately 25 percent less than for other medium-sized breeds.

The most common variety is White, but Crested ducks are also bred in Gray (Mallard color), Black, Blue and Buff. Hens lay blue, green, tinted or white eggs.

A trio of White Crested ducks. While none of these birds possesses large, round crests, they produced a good percentage of ducklings with head adornments that approached the ideal. Owned by Henry K. Miller.

Crested ducks are fast growing and highly decorative and the ducklings that hatch are strong and vigorous. Despite the disadvantage of having lower than normal hatchability, they are still a popular and practical breed.

Magpie. The name of this striking Welsh breed refers to its distinctive markings. The plumage is predominately white, offset by two colored areas that include the back (from the shoulders to the tail) and the crown of the head (see photo, page 34). The most common variety is Black, with a limited number of Blues and Duns also being raised. Breeding well-marked Magpies is a real challenge.

While Magpies are scarce, the few people raising them have found this duck to be productive and practical. Hens lay tinted or bluish-colored eggs and are among the best layers of the general purpose breeds. Magpies dress off cleaner than solid, dark-colored birds. This breed is an excellent duck for persons interested in helping to preserve a rare breed of waterfowl.

Orpington. William Cook of Kent, England, was the originator of the Orpington ducks. According to available records, they were introduced into North America in the early 1900s.

Five varieties of Orpingtons have been developed over the years—Buff, Blue, Silver, Black and Chocolate. The last four varieties all have white bibs on their chests. The Buff—which is solid colored—is the only variety com-

33

A Black and White Magpie duck. Ideally, black feathers do not extend down onto the thigh coverts, and bills are clear orange or yellow without green spots.

A Buff Orpington duck and drake possessing good length and depth of bodies, characteristics that are needed for heavy meat and egg production. Owned by Henry K. Miller.

monly raised today, and it is often referred to simply as the Buff duck. It is a pretty shade of fawn buff which does not become soiled as easily as white plumage, and yet is light enough that pin feathers are not unsightly when this variety is dressed for meat.

Orpington hens lay white or tinted eggs, and along with Magpies, are the best egg producers among the medium and large breeds. Many experienced waterfowl raisers consider the Buff Orpington the best choice when a single breed is desired that will provide both a steady supply of eating eggs and large roasting birds.

Swedish. Although their name would indicate otherwise, Swedish are believed to have been developed in Germany. They are bred in three varieties: Blue, Black and Silver. All varieties have white bibs that extend from under the bill to halfway down the breast. Ideally, the two outermost flight feathers on each wing are white.

While the color of the Blue variety is pretty, it is not the bright blue seen in some cage or wild birds, but rather is a subtle gray-blue shade. Blue Swedish hatch out only partially true to color, with approximately 50 percent of the ducklings being blue, 25 percent black and 25 percent silver or white splashed.

A Blue Swedish duck with typical color, but smaller than standard bib. Owned by Perrydale Acres, Amity, Oregon.

Hens lay blue, gray or tinted eggs. Swedish are extremely hardy, and are one of the most active foragers among the larger breeds. People who are intrigued by the principles of color inheritance find the Blues especially interesting to raise.

Bantam Breeds

These birds are the miniatures of the duck family and in the past have been valued primarily for their esthetic qualities. However, they are good spring layers if eggs are gathered daily, and their meat is exceptionally fine textured and has outstanding flavor. Highly adaptable, they thrive in limited space under close confinement, or can be allowed to roam at large. A patrol of several bantam ducks per acre of land provides an effective means of keeping insects and slugs under control. These breeds require a minimal quantity of grain or pelleted feed to supplement their gleanings. To obtain top egg production and hatchability (particularly under artificial incubation) it is often helpful to supply bantam ducks with a breeder ration containing 20 percent protein and fortified with 10 to 25 percent more vitamins than recommended for other breeds. Hens are excellent natural mothers.

Australian Spotted.   This handsome bird is seldom mentioned in poultry literature and infrequently advertised for sale. Nevertheless, they have been bred for at least fifty years, and are an established breed.

Apparently the Australian Spotted ducks kept in North America were created separately by several master waterfowl breeders. According to John C. Kriner, Jr. of Orefield, Pa., he and Stanley Mason developed this breed in the 1920s, and exhibited them for the first time in 1928. The late Henry K. Miller of Lebanon, Pa. was breeding and showing Spotted ducks of his own creation more than thirty-five years ago.

Typical body shape is intermediate between the long, racy wild Mallard and the short, compact Call. Coloration of the Australian Spotted varies, but in general, it is reminiscent of Mallards, with the striking difference being that the Australian drake displays conspicuous reddish-brown spots on his sides and back, a snowy chest and white flight feathers. The plumage of the female has an attractive frosted appearance with a liberal scattering of black or brown spots. Hens produce pale blue, greenish-buff or white eggs.

The Australian Spotted is a colorful and hardy little bird. It is often difficult to locate a source that offers stock for sale.

Call.   These tiny birds are the *smallest* of all domestic ducks, but make up for their lack of size with loud, persistent voices. In the era when live decoys were used in duck hunting, the English breeders were developing Calls for their talkativeness.

An Australian Spotted drake exhibiting well-marked sides. According to his owner, Henry K. Miller, this specimen should have a longer bill and a more streamlined head.

A White Call duck showing typical characteristics of short bill and horizontal carriage. She was photographed at the end of the breeding season when her head was not as large and round as normal. Owned by Henry K. Miller.

Plump, bowl-shaped bodies, short bills and large round heads are typical characteristics. Calls are bred in two common varieties: Gray and White. The Gray variety has the same color combinations as the Mallard, while the Whites are pure white with bright yellowish-orange bills, shanks and feet. Several rare varieties, including Blue, Buff, Golden, Aleutian and Spotted, are also being raised by waterfowl fanciers.

Call hens lay green, blue or tinted eggs. Newly hatched ducklings often are not as sturdy as the young of other breeds; therefore, some breeders have found it helpful to start them on a diet of quick-rolled oats, with finely chopped greens sprinkled on their water. When the ducklings are five to six days old, the oats can gradually be replaced with starting crumbles so that by two weeks of age, they are receiving 100 percent crumbles. Once started, Call ducklings are hardy and easily raised. The delightful, friendly Call duck is the most popular of the bantam breeds.

East Indie. It is believed that this breed originated in the United States, although English waterfowl breeders are given credit for its refinement. East Indies are usually a trifle larger than Calls, and exhibit a somewhat racier body conformation.

A group of East Indies with two old hens displaying white age feathering. Note the longer bills and racier type than typical for Calls. Owned by Henry K. Miller.

The iridescent green-black plumage of the Black East Indie cannot be described satisfactorily and live specimens must be seen in bright sunlight before their beauty can be fully appreciated. The brilliance of their feathers is *unequaled* by any other black domestic fowl. The same mating procedures are necessary as with the Cayugas to keep the occurrence of white feathers under control.

The first eggs of the laying season are black or dark gray, gradually changing to light gray or blue as the season advances. Hens are surprisingly good layers for such a small bird, and the ducklings are vigorous and hardy *if* they are out of a strain that is not overly inbred.

In our experience, East Indies reproduce extremely well under natural conditions with minimal care. If the ducklings are butchered at eight to ten weeks of age when in full-feather, they produce an attractive carcass that is just the right size to serve as a dinner for two.

Mallard. This wild species frequents most countries in the Northern Hemisphere, and is believed to be the parent stock of all domestic duck breeds, with the single exception of the Muscovy. Mallards readily adapt to domestication and in four or five generations of heavy feeding in captivity will in-

Mallard duck displaying her wing speculum, black and white wing-bars and intricately penciled body plumage. Authentic Mallard hens possess only one facial streak, while less typical birds often exhibit two. Owned by Dave and Millie Holderread.

Mallard drake in a relaxed pose with head feathers ruffled. He exhibits the long, slender bill, teardrop-shaped body and centrally placed legs typical of authentic specimens. Owned by Dave and Millie Holderread.

At seventeen weeks of age this Mallard drake has begun to replace his juvenile plumage with the bright colors of adulthood. To provide camouflage as long as possible, young males acquire their iridescent green heads and white neck collars last.

40

The same Mallard drake as shown to the left, photographed four months later in midsummer when he was in full eclipse plumage. By fall he will again sport the flashy nuptial plumage that makes the Mallard one of the most beautiful ducks.

crease dramatically in size and lose much of their desire and ability to fly. To preserve the streamlined type and outstanding foraging ability of the authentic Mallard, it is necessary to feed them judiciously and select breeders that most closely resemble the wild greenheads. (To legally capture wild Mallards or gather their eggs, a federal permit is required. However a permit is not needed to raise or sell domesticated Mallards.)

The Mallard drake is one of the *most beautiful* of all waterfowl with his iridescent green head, white neck collar, chestnut breast, silver underbody, blue wing speculum and reddish-orange feet. Except for her iridescent blue wing speculum, the hen's plumage is a mellow, though attractive, combination of almond and golden browns, with dark brown and black penciling. Besides the normal colored stock, there exist rare varieties that include the White, Snowy, Buff, Blue, Silver and Crested Mallards.

Along with Muscovies, genuine Mallards are the most self-reliant of all breeds. The hens are outstanding mothers and will hatch a high percentage of their greenish-buff or bluish eggs if not molested repeatedly while nesting. When there is an abundant supply of natural foods, hens are capable of raising their broods with little or no supplementary feed.

CHAPTER 6

Acquiring Stock

Having selected the breed or breeds you want to raise, the next step is locating suitable stock. The importance of starting with good quality birds cannot be overemphasized. The productivity, growth rate and size of ducks within the same breed varies a good deal. If your duck project is going to be economically practical and free of unnecessary problems, healthy and productive birds are essential.

PRODUCTION-BRED
VS. STANDARD-BRED STOCK

Egg and meat producing characteristics are given first priority in production-bred ducks, with less concern for perfect color or shape. On the other hand, standard-bred birds are used for showing in competition and are painstakingly selected for color, size and shape, with production abilities often given second priority. Normally, standard-bred stock is priced higher than production-bred stock.

In some breeds of ducks, the productivity and growth rate of exhibition strains are equal to or better than the commercial stock that is commonly sold by hatcheries. However, production-bred strains of Aylesburies, Rouens, Campbells and Runners are more practical for the home flock than are standard-bred birds of these breeds.

HATCHING EGGS

If a good dependable setting hen or incubator is available, you may wish to buy hatching eggs to start your flock. Some advantages of this method are that hatching eggs normally sell for one-third to one-half of the prices of day-old ducklings, and you get to experience the fun of waiting for and witnessing the hatch. Some disadvantages are that eggs vary in their fertility, they may be broken or internally damaged if they must be shipped, and it is impossible to know just how many ducklings will hatch.

If you receive a shipment of hatching eggs that are insured or COD for postage, open the package in the presence of your postal carrier to check for breakage and to count the number of eggs received. If a substantial number of eggs are broken or there are fewer eggs than you paid for, the postal carrier will provide a claim report.

Unless you know that the eggs are over two weeks old when you receive them, higher hatchability can often be obtained if shipped eggs are held at 55° to 65° F. for six to twelve hours prior to being placed in the incubator. Older eggs are best set promptly upon their arrival.

DAY-OLD DUCKLINGS

Purchasing day-old ducklings is the most popular method of starting a duck flock. They are more widely available than hatching eggs or adult stock. Ducklings are sturdy and can be shipped thousands of miles successfully. While living in Puerto Rico, we received hundreds of ducklings from distances of over 3000 miles, and the mortality rate was well under 1 percent.

Ducklings are occasionally sold sexed, but normally are available only straight run. Theoretically, unsexed ducklings run 50 percent drakes and 50 percent hens. Practically, there may be 10 to 25 percent more drakes than hens or vice versa, particularly when ducklings are purchased in lots of twenty-five or less.

When ordering ducklings, give your telephone number (or a neighbor's, if you don't have a phone) and instruct the shipper to include it on the shipping label. If you live on a long rural route, ask your local postmaster to hold the ducklings at the post office and phone you upon their arrival so you can pick them up promptly. When a shipment is received, open the box in the presence of the postal employee and check the condition of the ducklings and count the live birds. If the quantity of live ducklings you receive is significantly fewer than the number you paid for, the postal carrier should provide a claim report.

Day-old Khaki Campbell, Mallard and White Indian Runner ducklings several hours after being taken out of the incubator.

Care of Shipped Ducklings

The first twenty-four hours after ducklings arrive are *critical*. The birds should be given drinking water, food and rest as soon as possible. As you take the little ones from the shipping box and place them in the brooder, dip each of their bills in lukewarm water to help them locate the drinking fountain. They should be checked frequently the first day, but do not handle or disturb them more than absolutely necessary.

BUYING MATURE STOCK

The quickest way to obtain a producing duck flock is to purchase mature birds. Poultry farms and hobbyists often have adult stock available. Because the productivity of ducks normally decreases significantly after their third or fourth year, it is preferable to acquire birds that are not over one or two years of age.

When possible, it is advantageous to acquire ducks locally since you'll save on transportation costs, and the ducks will be seen at the time of purchase. If a problem arises at a later date, it will be convenient to communicate with the seller. However, waterfowl adapt quickly to new climates and are readily shipped, so you need not hesitate ordering from out-of-area breeders and hatcheries if the birds you want are not available locally.

Some good places to look for ducks or addresses of sources follow: feed stores; agriculture fairs; university poultry or animal science departments; agriculture extension services; classified ad sections of poultry, farm and gardening magazines, and local newspapers; and Appendix G (The Duck Breeders and Hatchery Guide) in this book.

HOW MANY

The ideal number of ducks for your flock will depend on the purpose you have for raising them, the breed raised and the environmental conditions to which the birds are exposed.

To estimate the number of hens needed for a laying flock, calculate the total number of eggs you will need over a year's time. Divide this number by the average number of eggs a hen of the breed you are going to raise will lay during a year. Keep in mind that the figures given in the Breed Profile Chart (page 21) for the yearly egg production of the various breeds are for hens that are fed concentrated feeds and exposed to no less than fourteen hours of light daily during the laying season. Egg production is reduced by 30 to 60 percent if concentrated laying feeds are not utilized and by 30 percent or more if artificial lighting is not used to lengthen the short days of fall, winter and spring. (See Table 18, page 105).

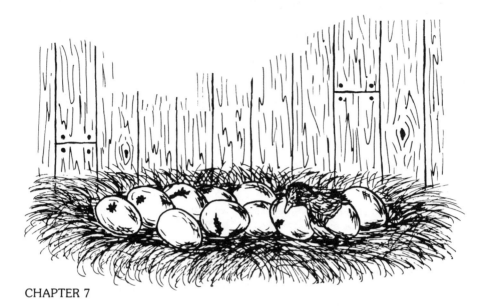

CHAPTER 7

Incubation

For many of us who own poultry, the incubation and hatching of eggs is the most fascinating phase of raising birds. When holding an egg in one's hand, it is difficult to comprehend that inside the shell there exists every element necessary for the beginning and growth of a new life. In fact, when a fertile egg is laid, an embryo several thousand cells in size has already formed. If stimulated by warmth and movement, that tiny spark of life will grow, break from the shell and present itself as a new generation of ducks!

HATCHING EGGS

Gathering

Eggs that are going to be incubated by a foster hen or in an incubator should be gathered at least twice daily to protect them from predators and prolonged exposure to the elements. Hatching eggs must *always* be handled gently so that the diminutive embryo is not injured or the protective shell cracked. Rolling eggs over and over, jolting them sharply or handling them with dirty hands—all can destroy fertility.

Cleaning

Eggs that are nest-clean hatch the best. Lightly soiled eggs should be wiped with a clean cloth or fine sandpaper. Dirty eggs need to be washed as soon after gathering as possible.

Washing has several negative effects on duck eggs: it removes the cuticle (a protective film on the shell that reduces dehydration and screens out pollutants); it can lower hatchability slightly; and it makes it necessary to raise the humidity level during incubation by 5 to 10 percent. Nonetheless, it is preferable to wash dirty eggs rather than set them uncleaned, since contaminated eggs create an unsanitary condition under the hen or in the incubator and frequently explode during incubation.

When eggs are washed, always use clean water that is 10° to 25°F. warmer than the eggs. Washing with dirty water spreads contaminants from egg to egg, while cold wash water forces filth deeper into the shell pores. In situations where it is practical, use a hatching egg disinfectant in the wash water.

Selecting

Not all fertile eggs are suitable for hatching. Those used for setting should have normal shells and be average to large in size. Extremely large eggs often have double yolks and seldom hatch. Any eggs having irregular characteristics or the slightest crack are best used for food. Valuable eggs with small cracks can sometimes be saved by placing a piece of tape over the fracture.

Storing

Proper care of eggs *prior to setting* is just as important as correct incubation procedures. In working with owners of small poultry flocks, I have found that careless handling of eggs before incubation is one of the leading causes of bad hatches. It must always be kept in mind that no matter how faithful a setting hen is or how carefully the incubator is regulated, a poor hatch will result if the embryos have been weakened or killed during the holding period.

Where. Eggs must be stored away from direct sunlight in a cool, humid location. Cellars and basements are ideal places; refrigerators are usually too cold.

Position. The position in which eggs are stored prior to incubation has little effect on hatchability. The results of a study involving thousands of duck and goose eggs showed no significant difference in the hatchability of eggs stored on their sides, vertical with the air cell up or vertical with the air cell down. However, when eggs are stored in twelve-egg cartons or thirty-egg flats (twenty-egg turkey flats are best for large duck eggs), less breakage occurs if the eggs are positioned with their large end up.

Temperature. The ideal storage temperature for hatching eggs that are held for ten days or less seems to be 55° to 65° F. If eggs are kept for a longer time, a temperature of 48° to 52° F. will produce better hatches. Because wide temperature fluctuations reduce the vitality of embryos, it is wise to store eggs where the temperature stays at a fairly constant level. In a study of the effects of storage temperature on hatchability in duck eggs that were stored five to seven days, the following results were obtained:

Storage Temperature	Hatchability Percentage
38°-40° F.	61
60°-62° F.	73
76°-82° F.	42

Turning. Duck eggs that are held five days or less show little improvement in hatchability when turned during the storage period. On the other hand, when eggs are stored longer than five days, the hatch can be increased 3 to 15 percent by turning them daily while they're being saved for incubation. If eggs are stored in egg cartons or flats, they can be turned by leaning one end of the container against a wall or on a block at an angle of 30° to 40° each day alternating the end that is raised.

Length of storage. Ordinarily, the shorter the storage period is, the better the hatch. A few eggs that have been held four weeks or longer may hatch, but for good results, the general rule is not to keep eggs for more than ten days before setting them. The negative effect of long storage periods on hatchability can be seen in the results obtained from a test involving several thousand eggs.

Length of Storage	Hatchability Percentage
1-7 days	71
8-14 days	64
15-21 days	47
22-28 days	18

Incubation Period

The normal incubation period for Mallards and their derivatives is twenty-seven to twenty-eight days, while Muscovies require approximately a week longer—thirty-three to thirty-five days. High temperatures during storage or incubation cause premature hatches, while long storage periods and low incubation temperatures result in late hatches.

Average Fertility

It is unusual for all eggs in a large setting to be fertile. The average fertility for heavyweight breeds is 85 to 95 percent and for lightweight birds, 90 to 98 percent. See Table 9, page 62, for common causes of poor fertility.

Average Hatchability

The hatchability of artificially incubated duck eggs often is 5 to 10 percent lower than for chicken eggs. Still, good setting hens frequently hatch every fertile duck egg they incubate. Under artificial incubation, the average hatchability falls between 65 to 85 percent of all eggs set, or 75 to 95 percent of the fertile eggs. See Table 9, page 62, for common causes of poor hatchability.

NATURAL INCUBATION

When you wish to hatch a moderate number of ducklings, natural incubation is often the most practical. A good setting hen is a master at supplying the precise temperature, and instinctively knows just how often eggs need to be turned. She also serves as a ready-made brooder, eliminating the need to supply an artificial source of heat.

Choosing Natural Mothers

The Breed Profile Chart (see page 21) under the heading of Mothering Ability, indicates the average capability of the various breeds as natural mothers. Duck eggs can also be hatched by goose, turkey and chicken hens. The breeds of chickens that make the best foster mothers are Silkie or common barnyard bantams and large Old English Games, Orpingtons and Cochins.

Clutch Size

Duck hens normally cover eight to fourteen (Muscovies sixteen to twenty) of their own eggs. Some hens lay such large clutches that they cannot incubate the eggs properly. In this situation, the oldest eggs—those that are the dirtiest—should be removed, leaving only the number that the hen can cover comfortably. Eggs must be positioned in a single layer to hatch well, never stacked on top of one another. If too many eggs are in a nest, the result will be a poor hatch or a complete loss.

Care of the Broody Hen

Setting hens are temperamental and should not be disturbed by people or animals. It is advantageous to isolate the broody from the rest of the flock with a temporary partition. This precaution will keep other hens from attempting to lay in the broody's nest and disrupting the incubation process. Unlike chickens, duck hens and their nests usually *cannot* be moved.

To remain healthy during her long vigil on the nest, the hen must eat a balanced diet, have clean drinking water, and be protected from the hot sun. Feed and water containers should be placed several feet from the nest so that the hen must get off to eat and drink. A leave of absence from the nest for five to ten minutes once or twice daily is essential to the hen's good health and will not harm the eggs.

When chicken or turkey hens are used to hatch duck eggs, they should be treated for lice and mites several days before their setting chores commence. It is usually necessary to sprinkle waterfowl eggs with lukewarm water several times each week when turkey or chicken hens are employed.

Multiple Broods

Muscovies—and occasionally hens of other breeds—will frequently bring off two broods a year or up to three in mild and tropical climates. Multiple broods can be encouraged by feeding hens extra feed in the early spring and by removing the ducklings at one to four weeks of age.

ARTIFICIAL INCUBATION

There are circumstances when an incubator is useful. Unlike setting hens, incubators can be used any season of the year, and come in such a wide range of sizes that any number of eggs, from one to thousands, can be set simultaneously or on alternate dates.

On the other hand, incubators must be attended to periodically each day to check the temperature and humidity, and to turn the eggs if this function is not performed automatically. The hatchability of eggs is lower and the number of crippled or weak young higher under artificial incubation than when the natural method is used. Also, electric incubators are at the mercy of power failures unless a gasoline generator is available.

Types of Incubators

Incubators are made and sold in a wide range of sizes and shapes, with varying degrees of automation. They can be divided into two basic types: the still-air (gravity flow) and the forced-air.

Still-air. These models, available with electric or oil heat, closely approximate natural conditions by having the heat source above the single layer of eggs, causing the upper surface of the eggs to be warmer than the lower portion. Still-air incubators are simple to operate, dependable, nearly maintenance-free and are manufactured with capacities of 20 to 400 eggs. We have used four models of still-air machines and have had excellent results with each. In my opinion, they are the best incubators for the small home poultry owner.

Forced-air. These incubators are equipped with fans or beaters which move warmed air to all surfaces of the eggs, and normally have multiple layers of egg trays. Forced-air machines are available with capacities of twelve to many thousands of eggs, and when compared to still-air incubators, are better suited to automatic turning of eggs, and take less floorspace for larger quantities of eggs. They are also more complicated, require greater maintenance and sell for higher prices.

Homemade. With a little ingenuity and a lot of persevering care, satisfactory hatches can be obtained in a homemade incubator consisting of a cardboard or wooden box and light bulbs for heat. More elaborate incubators, complete with heating elements and thermostats, can also be crafted in the home shop. (Kits for making small incubators are manufactured by the Lyon Electric Company. See Appendix H for the address.) In emergency situations —such as a hen deserting her nest—it is possible to hatch eggs in an electric frying pan or heating pad.

Where to Place the Incubator

Incubators perform best in rooms or buildings where the temperature does not fluctuate more than 5° to 10° F. over a twenty-four hour period. Consistent temperatures are especially important for still-air incubators, which should be located in a room with an average temperature of 60° to 70° F. *Do not* position your machine where it will be in direct sunlight, or near a window, heater or air conditioner.

Leveling the Incubator

Incubators, particularly still-air models, must be level to perform well. If the incubator is operated while askew, the temperature of the eggs will vary in different areas of the machine, causing eggs to hatch poorly and over an extended period of time.

Operating Specifications

Manufacturers of incubators include a manual of operating instructions with their machines. This guide should be carefully read and followed. The operating instructions often cannot be adapted from one machine to another with good results, particularly if one is a still-air model and the other a forced-air. If you acquire a used incubator which does not have an instruction booklet, manufacturers are usually willing to send a new manual if you send them a request with the model number of your machine.

The following is a summary of the basic incubation requirements of duck eggs.

Setting the eggs. Start the incubator *at least* forty-eight to seventy-two hours ahead of time and make all necessary adjustments of temperature, humidity and ventilation before the eggs are set. People frequently put eggs in machines that are not properly regulated, thinking they can make fine adjustments after the eggs are in place. This practice is a serious mistake since one of the most critical periods for the developing embryo is the first four or five days of incubation.

Prior to being placed in the incubator, duck eggs need to be warmed up for five or six hours at a room temperature of 70° F. If cold eggs are set without this warming period, water condenses on the shells, and yolks occasionally rupture.

For high percentage hatches, it is *essential* that eggs are incubated in the correct position. Always set them on their sides with the large end (air cell) slightly raised. When duck eggs are set with the air cell lowered, their chances of hatching are decreased by up to 75 percent. Set only the number of eggs that fits comfortably in the tray, without crowding or stacking them on top of one another. If at all possible, do not disturb eggs during the first twenty-four hours they are in the incubator.

Temperature. In still-air machines the correct temperature is 101.5° F., 102° F., 102.5° F., and 103° F. for each consecutive week. If you do not fill the incubator with one setting, but add a few eggs each week, a constant temperature of 102° to 102.5° F. should be maintained. It is essential that thermometers be positioned properly in still-air incubators or an incorrect temperature reading will be given. The top of the thermometer's bulb must be level with the top of the eggs. *Do not* lay the thermometer on top of the eggs since this practice will give a warmer temperature reading than actually exists at the level of the eggs. Forced-air machines are maintained at several degrees lower, 99.5° to 99.75° F., since all sides of the egg are warmed equally.

The temperature must be watched closely the last seven to ten days of incubation. During this period, an increase in temperature is often experienced, and the thermostat may need to be adjusted slightly each day to keep the eggs from overheating. Lowering the temperature by 1° to 1.5° F. for the final two days is beneficial, since ducklings generate considerable internal heat in their struggle to free themselves from their shells.

It is a good idea to use thermometers designed specifically for incubators, as they have greater accuracy than utility models and are easier to read.

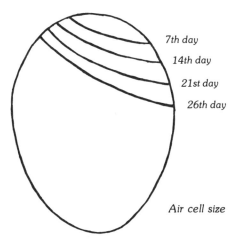

7th day
14th day
21st day
26th day

Air cell size

Humidity. To have a large number of strong ducklings hatch, the contents of the eggs must gradually dehydrate the correct amount. When dehydration is excessive, the embryos are puny and weak, making it extra tough for them to break out of the eggs. Conversely, inadequate moisture loss results in chubby embryos that have difficulty turning within the egg and cracking all the way around the shell.

The rate at which the contents of eggs dehydrate is regulated by the quantity of moisture in the air of the incubator. Moisture is usually supplied by water evaporation pans. To control humidity, the water surface area is increased or decreased, and the amount of ventilation regulated.

If your incubator is equipped with a wet-bulb thermometer or hygrometer, the correct reading on these instruments during the incubation period is 84° to 86° F., which is equal to a relative humidity of 55 percent. When eggs have been washed prior to incubation, the above figures may need to be raised to 88° to 90° F. on the wet-bulb thermometer or 65 percent on the hygrometer. Throughout the hatch (the last three days of incubation), the relative humidity should be increased to approximately 75 percent, or 92° to 94° F. on the wet-bulb thermometer.

While the hygrometer is a useful instrument for measuring the humidity level in incubators, the *best* indicator of whether the contents of the eggs are dehydrating at the correct rate is the size of their air cells. The air cell's volume can be observed by candling the eggs. On the seventh, fourteenth and twenty-first days of incubation, the average air cell volume should be approximately the same size as those in the accompanying illustration. If the air cells are too large, increase the moisture in the incubator by adding more water surface and/or decrease the amount of ventilation, being careful not to reduce the airflow so severely as to suffocate the embryos. If the air cells are too small, decrease the water surface area and/or increase the ventilation.

Due to variations in egg size and shell quality, the humidity requirements of duck eggs can vary considerably from one breed to another. Small eggs dehydrate faster than larger ones. When eggs from various breeds are incubated together, eggs from some breeds may dehydrate properly while others dry down excessively or not enough. In this situation, I have found it helpful if all the eggs from the breeds whose eggs dehydrate the slowest are washed prior to setting them, while the eggs from the breeds whose eggs dehydrate faster are not washed.

Ventilation. The developing embryo needs a constant supply of fresh air, which is provided through vent openings in the sides and tops of incubators. The amount of ventilation required is relatively small, but extremely essential to the well-being of the imprisoned ducklings. Ventilation demands increase at hatching time.

Turning. Incubating eggs must be turned to exercise the embryo and relieve stresses. Some eggs will hatch if you turn them just once every twenty-four hours, but turning three times daily at approximately eight-hour intervals is the *minimum* for high percentage hatches. For best results, duck eggs need to be rotated at regular hours and revolved at least one-third of the way around at each turning. Some studies indicate that turning eggs a full 180° improves hatchability. Eggs must be turned gently to avoid injury to the embryo. Turning should begin twenty-four to thirty-six hours after the eggs are set, and be discontinued three days before the scheduled hatch date.

When eggs are turned manually, it is helpful to mark them with an X and an O on opposite sides with a wax or lead pencil. Liquid inks—such as those in felt-tipped pens—should not be used, since they clog shell pores and can poison the embryos. After each turning, all eggs should have the same mark facing up.

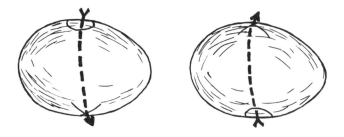

First turning is in this direction; next time, turn egg the opposite way.

Cooling. For best results when using still-air incubators, eggs should be cooled once daily—except during the first week and last three days of incubation. (Satisfactory hatches are obtained in forced-air machines without cooling.) When the room temperature is 65° to 70° F., the trays of eggs should be removed from the machine and cooled five minutes a day the second week, eight minutes daily the third week, and twelve minutes the first four days of the fourth week.

If you're like me, you will occasionally get sidetracked and forget the eggs as they cool. While they will hatch if left without heat for up to twelve hours once or twice during incubation (except during the first week and the last five days when low temperatures are disastrous), repeated over-cooling will retard growth and can be fatal. To avoid this problem, I borrow our kitchen timer and set it for the appropriate number of minutes at the beginning of each cooling period.

If you ever find that the temperature in the incubator is excessively high by more than 2° F., *immediately* cool the eggs for ten minutes and make adjustments to correct the problem.

Sprinkling. To prevent the egg membranes from drying out and becoming tough during the hatch, it is sometimes necessary to spray or sprinkle duck eggs with warm water forty-eight and again twenty-four hours before the calculated hatching time. While waterfowl breeders and incubator manufacturers frequently recommend that duck eggs also be wetted daily throughout the incubation period, good results can be obtained without this time-consuming chore if proper humidity levels are maintained throughout the incubation and hatching period. However, in tabletop incubators, the water evaporation pans often are so small that sufficient humidity cannot be maintained, and better hatching results in these machines are usually obtained when duck eggs are sprayed once daily from the 7th to 28th day of incubation. A convenient method is to use a discarded spray bottle that has been cleaned and filled with warm water.

Candling

The best time to candle duck eggs to check fertility is on the seventh day of incubation, unless they have dark-colored shells; then waiting until the tenth day is advantageous. If eggs are candled prematurely, it is more likely that fertile eggs will be accidently discarded.

Eggs are candled in a darkened room with an egg candler or flashlight. On the seventh day, fertile eggs reveal a small dark spot with a network of blood vessels branching out from it, closely resembling a spider in the center of its

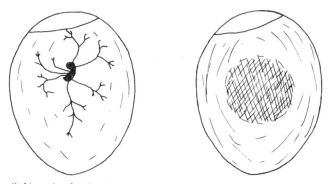

Fertile egg (left) and infertile egg (right) on the seventh day.

web. Infertile eggs are clear with the yolk appearing as a floating shadow when the egg is moved from side to side.

Sometimes embryos begin to develop, but perish within several days. When this happens, a streak or circle of blood is visible in an otherwise clear egg. Contaminated and rotten eggs often exhibit black spots on the inside of the shell, with darkened, cloudy areas floating in the egg's interior.

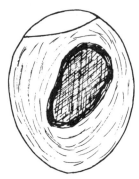

Blood ring — embryo has died.

All eggs not containing live embryos should be removed from the incubator when candled on the seventh or tenth day. Contaminated and rotting eggs give off harmful gases and frequently explode, covering the other eggs and the incubator's interior with putrid smelling goo that is difficult to clean up.

Uses for Infertile Eggs

Eggs that are infertile when candled on the seventh day of incubation are useful as a feed supplement for ducklings, adult poultry and other livestock such as pigs, cats and dogs. When fed to poultry or pets, eggs should *always* be hard-boiled and mixed with other feed. Feeding uncooked eggs can cause a serious biotin deficiency or result in birds and animals robbing nests to satisfy their appetite for raw eggs.

Eggs that have cloudy contents or dark blotches on the interior of the shell due to contamination are not suitable for livestock food. Rotten eggs should be disposed of where animals and birds will not have access to them.

THE HATCH

All the time and effort invested in producing and incubating eggs is rewarded by the hatch. Those first muffled chirps of the ducklings are sweet music.

Normally the first eggs will be pipped forty-eight hours prior to the hatch date. Ducklings require twenty-four to forty-eight hours to completely rim the shell and exit. Newly hatched birds are wet and exhausted, and should remain in the incubator four to twelve hours while gaining strength and drying off.

If your incubator is equipped with adjustable air vents, they should be regulated to give hatching ducklings additional air. However, do not open them so wide that an excessive amount of warm air escapes and the humidity drops. During the hatch, you may find it necessary to place extra water containers in the machine to maintain an adequate level of humidity with the increased air circulation. Evaporation pans should be covered with screen or hardware cloth to make *certain* that the ducklings cannot drown.

After most of the ducklings are hatched, the relative humidity can be lowered to 50 percent (82° to 84° F. on the wet-bulb thermometer) so they will fluff out properly.

Removing Ducklings From the Incubator

When the ducklings are dry, it is time to remove them from the incubator. Before opening the machine, have a clean container—with sides no less than six inches high—prepared with soft bedding. While transferring the ducklings, work quickly and gently, discarding empty shells and pipped eggs containing birds that are obviously dead. During this operation, the room temperature should not be below 60° F. Any ducklings that are still wet can be left in the incubator for several additional hours.

Help-Outs

At the end of the hatch a number of live ducklings may be still imprisoned within partially opened shells. Most of us find it hard to ignore these and wish to help them out. Assistance can be given by carefully breaking away the shell just enough so that the duckling will be able to exit under its own power.

While some birds that are assisted from the shell develop into fine specimens, a large percentage of them are usually handicapped by a deformity or
weakness. When it is understood that the hatch is a fitness test given by nature to cull out the weak and deformed—protecting them from facing a life for
which they are unprepared—we can take a more realistic view of helping
ducklings from the shell.

It is a good idea to mark all help-outs so they are not used as breeders unless they exhibit exceptional qualities as mature birds. Indiscriminate use of
such ducklings for breeding can lower the vitality of subsequent generations.
(See Marking Ducklings, page 88.)

Incubator Sanitation

While the incubator supplies the correct conditions for the embryo to develop, it also provides an excellent environment for the rapid growth of molds
and bacteria. Consequently, it is essential that the incubator be kept as clean
as possible.

At the conclusion of *each* hatch, clean and disinfect the incubator. If the
eggs are set to hatch at various times, the water pans should be emptied,
disinfected and returned with clean, warm water, and the duckling fuzz
removed from the incubator with a damp cloth or vacuum cleaner.

At the end of the hatching season, thoroughly clean the incubator and store
it in a dry, sanitary location.

FUMIGATION

On the average homestead, where relatively few eggs are set and hatched artificially each year, fumigation is normally not required for satisfactory hatching results. However, when eggs are set and hatched continuously over a period of time and the incubator cannot be thoroughly cleaned and disinfected
between each hatch, the bacteria count within the machine will soar, resulting
in reduced hatchability of the eggs and more deaths among the ducklings that
do hatch. When continuous setting and hatching of eggs is practiced, a sound
fumigation schedule will usually improve your results.

When to Fumigate

Table 8 gives a fumigation schedule. Of the five fumigations listed, numbers 1, 2 and 5 are usually sufficient. Number 1 is not as necessary if, promptly after they are gathered, eggs are washed with warm water to which a hatching disinfectant has been added.

TABLE 8
FUMIGATION SCHEDULE

Type of Fumigation	Amount of Fumigant per Cubic Ft. of Air Space		Length of Time Air Vents Should be Closed	Correct Conditions for Fumigation	
	Potassium-Permanganate	Formalin (37.5%)		Minimum Air Temp.	Minimum Humidity
1. Immediately after eggs are gathered and cleaned	0.6 grams	1.2 cc	20 minutes	70° F.	70%
2. Within 16 hours after eggs are set	0.4 grams	0.8 cc	20 minutes	99.5° F.	55%
3. 60 hours prior to end of incubation period	0.4 grams	0.8 cc	20 minutes	98° F.	65%
4. 30 hours prior to end of incubation period	none	0.5 cc	none	98° F.	65%
5. Empty incubators at beginning and end of season & between settings	0.6 grams	1.2 cc	3-12 hours	99° F.	55%

1. Fumigation number 1 should be carried out in a tight cabinet that has a fan for circulating the fumigation gas. Eggs should be placed on wire racks, in wire egg baskets or on clean egg flats. The sooner this fumigation takes place after the eggs are laid, the more effective it will be.

2. Fumigation number 2 takes place in the incubator within the first sixteen hours after the eggs are set, but not before the temperature and humidity have had a chance to normalize. Between the 24th and 120th hours of incubation, embryos go through a critical period. If eggs are exposed to formaldehyde gas during this time, the developing ducklings can be weakened or killed. However, when continuous setting is practiced in the same incubator, eggs can be fumigated several times during the course of the incubation period *if* they are not fumigated during this critical period.

3. When eggs are transferred to a separate machine for the hatch, fumigation number 3 can be employed to control outbreaks of omphalitis, a bacterial infection of the navel in newly hatched birds. To prevent damage to the lung tissue of the ducklings, fumigate *before* the eggs are pipped.

4. Fumigation number 4 is normally applied *only* when number 3 does not control omphalitis.

5. Prior to the first setting of the season, between hatches and at the end of the hatching season, fumigation number 5 should be used after incubators have been thoroughly washed and disinfected.

To be effective, a fumigation schedule should be designed and followed. Haphazard fumigating is of little value and is a waste of time and money. Under no circumstances should fumigation replace other sanitation procedures.

Procedures

The fumigation agent is formaldehyde gas which is produced by pouring liquid formalin over dry potassium-permanganate crystals. While formaldehyde gas is highly toxic, fumigation will destroy harmful bacteria and disease organisms without harming the eggs, the incubator or you, the operator, if the correct procedures are followed.

Step 1. Calculate the airspace within your incubator by multiplying the width times the depth times the height.

Step 2. Close all ventilation openings.

Step 3. Warm up incubator until it is operating at normal temperature and humidity levels for incubation.

Step 4. Measure out the correct quantity of potassium-permanganate into a dry earthenware, enamelware or glass container—*never* use metallic receptacles—having a capacity at least ten times the total volume of the ingredients, and place it in the incubator.

Step 5. Without leaning over the container, pour the correct amount of formalin over the potassium-permanganate and quickly shut the incubator door. The chemical reaction of these two substances begins within seconds. To prevent damage to your mucous membranes or burns on your skin, *never* combine these chemicals outside of the incubator or while holding the mixing vessel.

CAUTION

Formaldehyde gas is toxic to small and large organisms alike. When fumigating, follow instructions carefully and do not take chances that endanger your health.

Step 6. After the correct allotment of time, open the air vents to exhaust the formaldehyde gas. Thirty minutes to several hours after, remove the fumigation container and dispose of the residue in a safe location where children or animals cannot get into it.

If ducklings are fumigated thirty hours before the end of the incubation period (fumigation 4 in Table 8), no potassium-permanganate is used. Rather, sufficient cheesecloth is used to absorb the formalin and then hung in the incubator, allowing time for total evaporation. Position the cloth so that it *does not* touch the eggs or restrict air circulation.

INCUBATION CHECKLIST

1. Provide adequate nests furnished with clean nesting material.

2. Gather eggs in morning to protect them from temperature extremes.

3. Handle eggs with clean hands or gloves and avoid shaking, jolting or rolling them.

4. Store eggs in clean containers in a cool, humid location, and turn them daily if held longer than five days.

5. Set only reasonably clean eggs with strong shells.

6. Start incubator well in advance of using. Make adjustments of temperature, humidity and ventilation before eggs are set.

7. When filling the incubator, position eggs with their air cells slightly raised, and do not crowd eggs on tray.

8. Try not to disturb eggs during the first twenty-four hours they are in the incubator.

9. Gently turn eggs at least three times daily at regular intervals from the second to twenth-fifth day (second to thirty-second for Muscovies).

10. Operate still-air incubators at 101.5° to 103° F., and forced-air machines at 99.5° to 99.75° F. during the incubation period.

11. Maintain relative humidity at 55 percent (a wet-bulb reading of 84° to 86° F.) for first twenty-five days (thirty-two for Muscovies) of incubation.

12. When using a still-air machine, cool eggs five minutes a day the second week, eight minutes daily the third week, and twelve minutes the first four days of the fourth week.

TABLE 9: PINPOINTING INCUBATION PROBLEMS

Symptoms	Common Causes	Remedies
More than 10 or 15 percent clear eggs when candled on seventh day of incubation	1) Too few or too many drakes 2) Old, crippled or fat breeders 3) Immature breeders 4) Breeders frequently disturbed 5) No swimming water for mating 6) First eggs of the season 7) Late season eggs 8) Medication in feed or water 9) Eggs stored more than ten days	1) Correct drake to hen ratio 2) Young, active, semi-fat breeders 3) Use breeders seven months or older 4) Work calmly around breeders 5) Swimming water for large breeds 6) Don't set eggs laid first week 7) Don't set eggs when males molt 8) Avoid medicating breeders 9) Set fresher eggs
Blood rings on seventh to tenth day	1) Faulty storage of eggs 2) Irregular incubation temperature	1) Proper storage prior to setting 2) Adjust machine ahead of time
Ruptured air cell	1) Rough handling or a deformity	1) Handle and turn eggs gently
Yolk stuck to shell interior	1) Old eggs which haven't been turned regularly during storage	1) Turn eggs daily if they are held for more than five days
Dark blotches on shell interior	1) Dirt or bacteria on shells causing contamination of the inner egg	1) Wash eggs soon after gathering with water and disinfectant
More than 5 percent dead embryos between seventh and twenty-fifth day of incubation	1) Inadequate breeder diet 2) Highly inbred breeding stock 3) Incorrect incubation temperature 4) Periods of low or high temperature 5) Faulty turning during incubation	1) Supply balanced diet 2) Introduce new birds to flock 3) Check accuracy and position of thermometer 4) Check temperature frequently; don't overcool eggs 5) Turn at least three times daily
Early hatches	1) High incubation temperature	1) Lower incubation temperature
Late hatches	1) Low incubation temperature	1) Raise incubation temperature
Eggs pip but do not hatch; many fully developed ducklings dead in shells that aren't pipped	1) High humidity during incubation 2) Low humidity during incubation 3) Eggs chilled or overheated during last five days of incubation 4) Low humidity during the hatch causing egg membrane to dry out 5) Poor ventilation during hatch 6) Disturbances during hatch	1) Decrease amount of moisture 2) Increase amount of moisture 3) Protect eggs from temperature extremes during this period 4) Last three days, raise humidity to 75% and sprinkle eggs daily 5) Increase air-flow during hatch 6) Leave incubator closed
Eggs pipped in small end	1) Eggs incubated in wrong position	1) Position eggs with small end lower than the large end
Sticky ducklings	1) Probably low humidity during incubation and/or hatch	1) Increase amount of moisture
Large, protruding navels	1) High temperature 2) Excessive dehydration of eggs 3) Bacteria infection	1) Lower temperature; check thermometer 2) Increase humidity level 3) Improve sanitation practices
Dead ducklings in incubator	1) Suffocation or overheating	1) Increase ventilation during the hatch and watch temperature carefully
Spraddled legs	1) Smooth incubator trays	1) Cover trays with hardware cloth
More than 5 percent cripples other than spraddled legs	1) Inadequate turning during incubation 2) Prolonged periods of cooling 3) Inherited defects	1) Turn eggs a minimum of three times daily 2) Don't forget eggs when cooling 3) Select breeders free of defects

13. Lower the incubator temperature 1° to 1.5° F. for the hatch.

14. Increase the relative humidity to 75 percent (a wet-bulb reading of 90° to 94° F.) for the hatch.

15. Sprinkle eggs with lukewarm water forty-eight and twenty-four hours prior to the scheduled hatch date.

16. Do not open incubator except when necessary during the hatch.

17. Leave ducklings in machine until they are dried.

18. Clean and disinfect incubator after each hatch.

CHAPTER 8

Rearing Ducklings

The downy young of all types of poultry are charming, and day-old ducklings are no exception with their bright eyes, tiny wings and miniature webbed feet. Aided by shovel-like bills, they are soon eating an amazing quantity of food, all the while growing at an equally astounding rate. In only eight to twelve weeks, a newly hatched duckling is transformed into an adult duck.

BASIC GUIDELINES

Of all domestic fowl, the young of ducks are the easiest to raise. It is common for every duckling in a brood to be reared to adulthood without a single mortality. They withstand less than optimum conditions well—although that is no reason to mistreat or neglect them. If you put into practice the following management guidelines, raising ducklings will be a pleasant and trouble-free task.

Keep them warm and dry and protect them from drafts. Ducklings that are cold lose their appetites, and when wet, they chill rapidly and will die if not dried and warmed promptly. Drinking water containers should be designed so that ducklings cannot enter and become excessively wet.

Maintain them on dry bedding that provides good footing. Many internal parasites, molds and disease organisms thrive in damp and filthy bedding. Smooth, slick floors are the leading cause of spraddled legs.

Supply fresh, nonmedicated feed that provides a balanced diet. A proper diet is essential for disease resistance and fast growth. Feed that is moldy or contains medication can cause stunted growth, sickness and death. (See Medication Poisoning, page 126.)

Provide a constant supply of fresh water. Ducklings suffer a great deal when drinking water is not available, particularly right after they have eaten.

Furnish adequate floor space and fresh air. Forcing ducklings to live in crowded conditions is one of the leading causes of feather eating and disease.

Protect them from predators. Tame and wild predatory animals and birds find unprotected ducklings easy prey. The clumsy feet of humans and large animals also snuff out the lives of many young birds.

NATURAL BROODING

For the home poultry flock, natural brooding has many advantages. It eliminates purchasing or constructing special brooding equipment and supplying artificial heat. When allowed to range, hen-mothered ducklings learn to forage for their food at a young age, and will be nearly self-sufficient if there is a plentiful supply of insects, tender grass and wild seeds and fruits. Ducklings can be brooded by duck, turkey, bantam or large chicken hens.

Small brooder house with wire-covered run

A day-old Mallard
exhibiting the bold black
and yellow markings that
are typical of all gray
varieties of ducklings.

By two weeks of age, the
down color has faded and
feathers can be felt protruding from her tail and sides.

Managing the Hen and Her Brood

The best management practice in dealing with a hen and her brood of ducklings is to bother them as *little* as possible. Your main concern is to protect them from predators and to keep the ducklings from becoming soaked during their first several weeks of life outside the shell.

Free-roaming turkeys and chickens often leave ducklings stranded, so it is a good idea to pen these foster hens with their broods in a dry building or pen for the first two weeks. There the entire family will be protected from rain, cold winds, predators and rodents, and concentrated feed can be supplied to help them get off to a fast start.

The same procedures are recommended for duck hens and their broods, although it is normally safe to give them freedom in a large pen not occupied by other fowl after the ducklings are several days old. Duck hens are not as likely to jump or fly over barriers such as fences and leave the little ones behind.

Confine hens and their broods in a secure pen or building each night until

At four weeks, feathers cover the face, shoulders, underbody and tail, and have begun to appear on her back.

Now eight weeks old, the once soft, cuddly duckling is fully feathered and nearly the size of her parents.

the ducklings are six to eight weeks old. This significantly reduces the possibility of losses to predators.

Hens can be aggressive towards strange ducklings and may injure or kill them. Don't confine two or more hens together with their broods, particularly during the first week or two.

Allowing a new brood in with an established flock can also be dangerous for the young ones, especially when ducks are penned in close confinement and the strange ducklings may threaten the territorial boundaries of the adults. If your flock of ducks is permitted to roam freely, ducklings may not be bothered. In any case, newly hatched ducklings should be allowed with adult birds *only* if you can be on hand to remove them if a problem develops.

Giving Hens Foster Ducklings

Sometimes it is desirable to give hens foster ducklings to brood. If a hen already has young, the foster ducklings should be approximately the same size and color as her own. Hens who set, but for one reason or another do not hatch any ducklings—or chicks in the case of a chicken hen—often will accept a foster brood. Giving ducklings to a chicken hen that has chicks of her own usually does not work out, but it can be attempted in an emergency situation. To lessen the possibility of rejection, it is advisable to slip foster ducklings under their new mother after dark.

Novice duck raisers sometimes lock a hen (which has been setting for a short time, or not at all) in a pen with ducklings and expect her to mother them. Almost without exception these attempts fail and frequently result in ducklings being brutally attacked. Unless an exceptional hen is discovered, it is not wise to give foster ducklings to hens that have not set their full term.

The number of ducklings a hen can brood depends on her size and the weather conditions. Typically, a duck hen can successfully brood eight to twelve of her own young. Bantam hens can handle six to eight ducklings, large chickens twelve to eighteen, and turkey hens up to twenty-five. The most important factor is that all the ducklings can be hovered and warmed at the same time.

Excellent natural mothers, Muscovy ducks normally bring off large broods.

ARTIFICIAL BROODING

If a broody hen is not available or you are raising a large number of birds, it will be necessary to brood ducklings artificially. When sound management practices are used, this method produces good results.

The Brooder

Brooders provide warmth for the ducklings during the first four to six weeks after they hatch. Brooding equipment can be purchased from stores handling poultry supplies or can be fabricated at home with a minimum of skill and cost.

Battery brooder. This brooder is an all-metal cage which is equipped with a wire floor, removable dropping pan, thermostat, electric heating element, and feed and water troughs. They can be purchased and used in individual units or stacked on top of one another. Battery brooders are commonly used for starting chicks, but can also be used for brooding ducklings up to an age of two to six weeks, depending on the breed. While new units are expensive, this type of brooder requires limited floor space, is easily cleaned, protects young birds from predators and provides sanitary conditions.

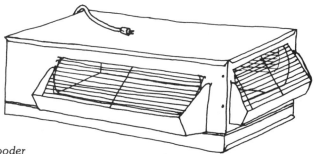

Battery brooder

Hover brooder. Available with gas, oil or electric heat, this brooder consists of a thermostatically controlled heater which is covered with a canopy. The brooder is supported with adjustable legs or suspended from the ceiling with a rope and is easily set to the proper height as the birds grow. Water fountains and feeders are placed around the outside of the canopy where the ducklings drink, eat and exercise in cooler temperatures. Hover brooders are available in sizes which are rated at 100 to 1000 chick capacity. The number

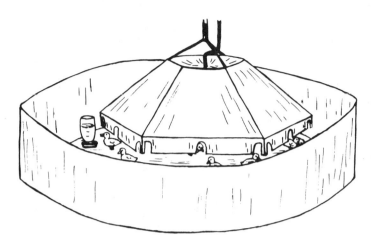

Hover brooder and draft guard

of ducklings under each unit should be limited to 50 to 60 percent of the given capacity for chicks.

Heat lamp. A simple method for brooding ducklings is with 250-watt heat lamps. Suspended eighteen to twenty-four inches above the litter, each lamp will provide adequate heat for twenty to forty ducklings, depending upon the size of birds and the outside temperature. We always use at least two lamps just in case one burns out.

When heat lamps are used, *extreme care* must be taken to prevent fires. These lamps must *never* be hung with the bottom of the bulb closer than eighteen inches from the litter, or so that any part of the bulb is near flammable material such as wood or cardboard.

Homemade brooder. A few ducklings can be brooded with a light bulb that is positioned in a box or wire cage. Recent research indicates that blue bulbs are best since they reduce the incidence of feather eating and are gentler on the ducklings' eyes.

The wattage required depends on the size of the box and the room temperature. I prefer to use several forty-watt bulbs rather than a single larger one. The box or cage must be big enough to allow ducklings to move away from the heat when they desire. When bulbs of over forty watts are used, they must be located out of the reach of ducklings. To prevent fires or the asphyxiation of ducklings from smoke produced by smoldering materials, even bulbs of low wattage *must not touch* or be close to flammable substances.

A light reflector and clamp make a safe and inexpensive "hover brooder." They are available at hardware stores and, while not used for brooding, come in handy around the shop and house. The reflector can be clamped to the side of a cage or box and will brood up to a dozen ducklings. As the young ones grow, the heat source can be adjusted to the correct height.

A variety of homemade brooders with more permanent qualities can also be crafted. Hover-type brooders can be built with a plywood or sheet-metal canopy and porcelain light fixtures. Better still is the washtub brooder built and used by a retired coal miner friend of ours. He outfitted an old zinc washtub with several light fixtures and three adjustable legs. With just a little work and a lot of ingenuity, he had a safe brooder that gave dependable service for many years.

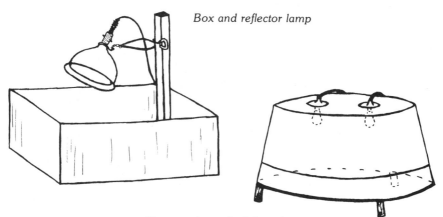

Box and reflector lamp

Homemade washtub brooder

Draft guard. The use of a draft guard around hover brooders and heat lamps is recommended for the first week or two. The guard protects ducklings from harmful drafts and prevents them from wandering too far from the heat and piling up in corners. Commercially produced guards constructed of corrugated cardboard can be purchased and used several times. Homemade guards can be fashioned with twelve-inch boards or welded wire that is covered with burlap or paper feed sacks. The guard should form a circle two or three feet from the outside edge of the hover brooder canopy.

Heat

Ducklings require less heat than chicks. Under the brooder the temperature should be held at approximately 90° F. the first seven days, and then lowered 5° each successive week. Once ducklings are six to eight weeks old and well-feathered, they can withstand temperatures down to 50° F. or lower, but must be protected from *drastic* temperature fluctuations.

The actions of the ducklings are a better guide to the correct temperature than a thermometer. If ducklings are noisy and huddle together under the heat source, they are cold and additional heat should be supplied. When they stay away from the heat, or pant, they are too warm and the temperature needs to be lowered. The proper amount of heat is being provided when ducklings sleep peacefully under the brooder or move about freely, eating and drinking.

Even at the start of the brooding period, it is *extremely* important that ducklings are able to get away from the heat source when they desire. Overheating is almost as damaging to ducklings as chilling.

Brooding Without Artificial Heat

If you're in an area not served by electricity or have chosen to live without it, you can still brood ducklings. One technique is to keep ducklings in a box near the stove or furnace until they are large enough to be put outside. If you use this method, be careful not to place the container too close to the heat, which could overheat the ducklings or, worse yet, start a fire.

Another procedure that works satisfactorily in mild weather is to utilize the body heat given off by the ducklings to warm themselves. A well-insulated box is the basic equipment needed. The floor of the container should be covered with two to four inches of dry bedding such as clean rags, chopped grass, straw or sawdust. If a fine bedding such as sawdust is used, cover it with cloth or burlap the first several days to prevent ducklings from harming themselves by eating the particles of wood.

The top and sides of the box must be draped with a layer of old towels, blankets, or burlap bags. Be sure that the little ones have sufficient air to prevent suffocation. Since heat rises, the ducklings will stay warmer if the "hot box" you design has a maximum of six inches of headroom after the bedding is in place. Also keep in mind that the smaller the inside area is, the cozier the occupants will be.

A small door should be made in the side of the brooder box where the ducklings can exit to eat and drink. They soon learn to go back inside when they're chilly, although you'll probably need to give them a helping hand the first day or two.

FLOOR SPACE

When ducklings are started on litter, allow a minimum of .75 square feet of floor space per bird for the first two weeks, 1.75 square feet until four weeks of age, 2.75 square feet until six weeks, and 3 to 5 square feet per bird thereafter. At three to four weeks of age, give ducklings access to an outside pen or yard during mild weather, allowing 10 square feet of space per bird. This additional space will help keep the inside bedding drier and reduce sanitation problems.

LITTER

Ducklings consume a huge volume of water so it is *exceedingly important* that a thick layer of absorbent, mold-free litter (or wire flooring) is used to keep the brooding area from becoming sloppy. Shavings, sawdust, peanut hulls, peat moss, crushed corncobs, flax or chopped straw can be used for bedding.

Begin with three to six inches and add new litter as required. Soggy and caked bedding should be removed whenever it appears and replaced with dry material. A daily stirring of litter is advantageous, particularly if the density of ducklings is high. In warm weather, flies will become a problem if the quality of the litter is allowed to deteriorate.

YARDS AND PASTURES

Ducklings can be given access to a yard or pasture as soon as they are comfortable outside. The exercise, fresh air and sunlight are beneficial to their health. They will also eat substantial quantities of tender grass which reduces their feed consumption and enriches their diets with vitamins and minerals. Ducklings *cannot* eat mature, dry or coarse grasses.

When allowed outside, ducklings should be put under cover each time it rains until they are five to six weeks old. After this age, a simple shelter is adequate protection.

WATER

Swimming Water

It is not necessary to have swimming water for ducklings, even though they thoroughly enjoy going for a paddle within hours after hatching. Even when ducklings are brooded by a duck hen, it is safest to keep them out of water until they are at least two weeks old.

To protect ducklings from drowning, all water containers which they can enter should have gently sloping sides with good footing to allow tired and wet swimmers to exit easily. If you supply swimming water in receptacles having steep or slick sides, an exit ramp *must* be provided if drowning losses are to be avoided. Drowning and becoming soaked and chilled while swimming are the leading causes of duckling mortalities in home duck flocks.

Aylesbury, Pekin, Rouen and White Call ducklings of various ages enjoying a small pond made by mortering together a row of cinder blocks across a creek on Boyd Smith's farm in Kunkletown, Pennsylvania.

Drinking Water

To thrive, ducklings must have a *constant* supply of drinking water. Drinking fountains should be designed so that young ducklings *cannot* get into the water. To lessen the possibility of ducklings choking to death, the water should be deep enough to allow them to submerge their bills and dislodge particles of food which frequently become stuck in their throats and nostrils.

Sufficient receptacles should be provided so their contents are not quickly exhausted and the ducklings left without water. Placing water containers on screen-covered platforms is a big help in keeping the watering area dry and sanitary. Waterers should be rinsed out daily.

NUTRITION

The importance of a *sound* feeding program cannot be overemphasized. Nutrition is probably the most neglected phase of management in many small home flocks. Nearly half of the small duck flocks I observe exhibit nutritional deficiencies that can be avoided if the following recommendations are implemented.

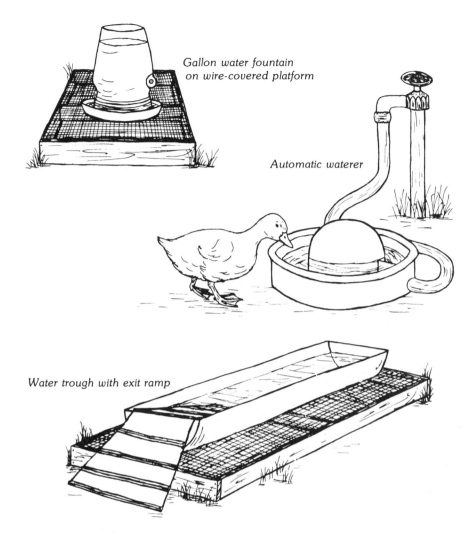

Gallon water fountain
on wire-covered platform

Automatic waterer

Water trough with exit ramp

The rate at which ducklings grow is in direct proportion to the *quantity* and *quality* of the feed they consume. For maximum growth they need a diet that provides 20 to 22 percent protein up to two weeks of age, and 16 to 18 percent protein from two to twelve weeks.

To stimulate fast growth, ducklings should be allowed to eat all the feed they want up to two weeks of age. After this time, you can limit them to two or three feedings daily, when they should be given all the feed they can clean up in five to ten minutes. When month-old ducklings have access to succulent pasture, they can be limited to one feeding daily. Giving the birds their meal in the evenings will encourage them to forage throughout the day.

For the first several weeks, small pelleted (3/32″) or coarse crumbled feed is preferred; thereafter, larger pellets (3/16″) will give the best results. Ducklings choke on fine, powdery mash when fed dry and up to 25 percent of the feed is wasted. If used, finely ground feed is better utilized when it is moistened with water or milk to a consistency that will form a crumbly ball when compressed in your hand. A new batch should be mixed up at each feeding to avoid spoilage and food poisoning.

TABLE 10
SUGGESTED FEEDING SCHEDULE FOR DUCKLINGS

Type of Duckling	0-2 Weeks	2-7 Weeks	7-20 Weeks
	Pounds of 20% Starter Feed per Bird Daily	Pounds of 18% Grower Feed per Bird Daily	Pounds of 16% Developer Feed per Bird Daily
"Green"	Free choice	Free choice	
Small Breed	Free choice	Free choice 5 min. twice daily	0.15-0.25
Egg Breed	Free choice	Same as above	0.20-0.30
Medium Breed	Free choice	Same as above	0.25-0.35
Large Breed	Free choice	Same as above	0.30-0.40
Muscovy	Free choice	Same as above	0.20-0.40

The quantity of feed required by ducklings is highly dependent on the availability of natural foods, climatic conditions and the quality of feed (e.g., birds require larger amounts of high fiber foods than low fiber foods to meet their energy requirements).

Feeding Programs

The feed program you employ should be designed to fit your situation and goals, and most likely will be a combination of two or more of the following options.

Natural. Ducklings, particularly Muscovies and Mallards, which are brooded by their natural mothers, are capable of foraging for most of their own ration if there is an abundant supply of insects, wild seeds and succulent plants. Ponds, lakes, sloughs, marshes and slow-moving brooks are excellent sources of free food for ducklings. In most situations it is advisable to supply ducklings concentrated feed for at least the first ten to fourteen days to get them off to a good start.

Grains. Small whole grains such as wheat, milo, kafir or cracked corn can be fed to ducklings after they are several weeks old. Grains by themselves are *not* a balanced diet. Ducklings need tender greens and an abundant supply of insects or another protein supplement. They cannot be expected to remain healthy and grow well on a diet consisting exclusively of whole, cracked or rolled grains.

Home Mixed. It may be practical to mix your ducks' feed if the various ingredients are available at a reasonable price. The formulas given in Tables 11 and 12 are examples of the types of feed that can be mixed at home. While these rations are not as sophisticated as commercially prepared feeds, they will give good results in most situations *if* they are mixed properly and the instructions are closely followed. If a vitamin:mineral premix (carried by many feedstores) is used, the rations in Table 13 can also be home-blended. For instructions on how to formulate rations using locally available foodstuffs, see Appendix A, page 135.

Special equipment is not needed to mix duck feed. For small quantities the ingredients can be placed in a large tub and combined with the hands or a stick. Another method is to pile the measured components in layers on a clean floor and mix with a shovel. A cement mixer or old barrel mounted on a stand and outfitted with a handle, door and ball bearings can be used for larger quantities. More important than the method used for mixing is that the ingredients are blended *thoroughly*. During warm weather, no more than a four-week (three weeks or less if ground grains are used) supply of feed should be prepared at a time.

TABLE 11
HOME-MIXED STARTING RATIONS (0 to 2 WEEKS)

Ingredient	No. 1 Small Quantity		No. 2 Large Quantity	
Yellow cornmeal	11	cups	62	lbs.
Soybean meal (44% protein)	3½	cups	17	lbs.
Wheat bran	2	cups	2	lbs.
Meat and bone meal (50% protein)	½	cup	4	lbs.
Fish meal (60% protein)*	½	cup	2	lbs.
Alfalfa meal (17.5% protein)	½	cup	2	lbs.
Dried skim milk or calf manna	½	cup	3	lbs.
Brewer's dried yeast	1½	cups	7¼	lbs.
Dicalcium phosphate (18.5% P)	1	tbsp.	½	lb.
Iodized salt	1	tsp.	¼	lb.
Totals	20	cups	100	lbs.
Cod liver oil**				
Chopped succulent greens	Free choice		Free choice	
Sand or chick-sized granite grit	Free choice		Free choice	
Chick-sized oyster shells or				
crushed dried egg shells	Free choice		Free choice	

* If fish meal isn't available, use 4½ cups soybean meal and 10½ cups cornmeal in formula #1, or 20 lbs. soybean meal and 61 lbs. cornmeal in formula #2.

** If birds do not receive direct sunlight, which enables them to synthesize vitamin D, sufficient cod liver oil must be added to these rations to provide 500 International Chick Units (ICU) of vitamin D_3 per pound of feed.

TABLE 12
HOME-MIXED GROWING RATIONS (2 to 12 WEEKS)

Ingredient	No. 3 Corn Base (pounds)	No. 4 Wheat Base (pounds)	No. 5 Milo Base (pounds)
Cracked yellow corn*	75.00	—	—
Whole soft wheat	—	79.00	—
Milo (grain sorghum)	—	—	76.00
Soybean meal (50% protein)	9.00	5.00	8.00
Meat and bone meal (50% protein)	4.00	4.00	4.00
Alfalfa meal (17.5% protein)	3.00	3.00	3.00
Dried skim milk	3.00	3.00	3.00
Brewer's dried yeast	5.00	5.00	5.00
Dicalcium phosphate (18.5% P)	0.10	0.10	0.10
Limestone flour or oyster shells	0.65	0.65	0.65
Iodized salt	0.25	0.25	0.25
Total (lbs.)	100.00	100.00	100.00
Cod liver oil**			
Sand or chick-sized granite grit	Free choice	Free choice	Free choice
Chopped succulent greens (eliminate when pasture is available)	Free choice	Free choice	Free choice

* Whole corn can be used after the birds are 4 to 6 weeks of age.
** See Table 11.

Commercial. In some localities premixed starter and grower feeds for ducklings are available. When these feeds are used, the instructions should be followed. If rations specifically formulated for ducklings are not available, use the corresponding nonmedicated mixes recommended for game birds, turkeys or chickens. The first two are preferred, but chicken feeds usually give satisfactory results if they are fortified with additional niacin.

You may want to have a local feed mill mix your feed if enough ducklings are raised to make it practical. The formulas given in Table 13 are for complete rations that will provide a balanced diet *if* mixed properly and not stored for more than four weeks (less in hot weather). Ration numbers 6, 7, and 8 are to be fed to ducklings up to two weeks of age, and numbers 9, 10 and 11 from two to twelve weeks.

Niacin Requirements

Young waterfowl require two or three times more niacin in their diet than chicks. (See Niacin Deficiency, page 126.) When ducklings are raised in confinement on a ration that is deficient in niacin—such as commercial chick feeds—a niacin supplement should be added to their feed or water.

TABLE 13
COMPLETE RATIONS FOR DUCKLINGS (PELLETED)

Ingredient	No. 6 Corn Base Starter (lbs/ton)	No. 7 Wheat Base Starter (lbs/ton)	No. 8 Milo Base Starter (lbs/ton)	No. 9 Corn* Base Grower (lbs/ton)	No. 10 Wheat* Base Grower (lbs/ton)	No. 11 Milo* Base Grower (lbs/ton)
Ground yellow corn	1385	—	—	1570	—	—
Ground soft wheat	—	1425	—	—	1607	—
Ground milo (grain sorghum)	—	—	1377	—	—	1575
Soybean meal solv. (50% protein)	430	360	425	295	270	330
Fish meal (60% protein)	40**	40**	40**	—	—	—
Meat and bone meal (50% protein)	80	80	80	80	40	40
DL-Methionine (98%)	2.00	2.75	3.25	1.50	2.00	2.50
Stabilized animal fat	25	40	30	—	20	—
Soybean oil	—	14	5	—	10	—
Dicalcium phosphate (18.5% P)	6.25	5.25	6.25	8.00	17.00	18.00
Limestone flour (38% Ca)	6.00	7.00	7.75	20.00	9.00	9.50
Iodized salt	5.75	6.00	5.75	5.50	5.00	5.00
Vitamin:mineral prefix	20	20	20	20	20	20
Totals (lbs.)	2000	2000	2000	2000	2000	2000

Vitamin:Mineral Premix

Vit. A (millions of IU/ton)	8.0	9.5	9.5	5.0	6.5	6.5
Vit. D$_3$ (millions of ICU/ton)	1.0	1.0	1.0	.8	.8	.8
Vit. E (thousands of IU/ton)	5.0	15.0	15.0	2.0	12.0	12.0
Vit. K (gm/ton)	2.0	2.0	2.0	2.0	2.0	2.0
Riboflavin (gm/ton)	6.0	6.0	6.0	4.0	4.0	4.0
Vit. B$_{12}$ (mg/ton)	8.0	8.0	8.0	4.0	4.0	4.0
Niacin (gm/ton)	50.0	50.0	50.0	40.0	40.0	40.0
d-Calcium pantothenate (gm/ton)	6.0	3.0	3.0	4.0	2.0	2.0
Choline chloride (gm/ton)	300.0	—	300.0	200.0	—	200.0
Folic acid (gm/ton)	.5	.5	.5	.5	.5	.5
Manganese sulfate (oz/ton)	9.6	9.6	9.6	9.6	9.6	9.6
Zinc oxide (80% Zn, oz/ton)	3.2	3.0	3.2	3.2	3.0	3.2
Ground grain to make 20 lbs.	+	+	+	+	+	+
Totals (lbs.)	20.0	20.0	20.0	20.0	20.0	20.0

Calculated Analysis

Crude protein (%)	20.2	19.6	20.1	16.4	16.1	16.4
Lysine (%)	1.05	1.00	1.03	.76	.73	.75
Methionine (%)	.46	.44	.46	.37	.33	.35
Methionine + cystine (%)	.76	.76	.76	.62	.63	.61
Metabolizable energy (kcal/lb)	1424	1366	1416	1426	1362	1410
Calorie:protein ratio	70	70	70	87	85	86
Crude fat (%)	4.7	4.7	4.2	3.5	3.3	2.6
Crude fiber (%)	2.3	2.4	2.3	2.3	2.4	2.3
Calcium (%)	.76	.78	.80	.93	.63	.64
Available phosphorus (%)	.41	.41	.42	.36	.36	.37
Vit. A (IU/lb)	4912	4750	4750	3535	3250	3250
Vit. D$_3$ (ICU/lb)	500	500	500	400	400	400
Riboflavin (mg/lb)	3.73	3.76	3.76	2.64	2.66	2.66
Total niacin (mg/lb)	41	51	46	31	43	37

* To convert into breeder-developer rations, substitute the following quantities of vitamins for those listed under the vitamin:mineral premix: 1 gm of vit. K; 3 gms riboflavin; 30 gms niacin; and 100 gms choline chloride in the corn and milo base rations.

** If fish meal is not available, use an additional 50 lbs soybean meal solvent (50% protein) and subtract 10 lbs of grain.

KEY: IU = International Units   gm = gram   kcal = kilocalories
     ICU = International Chick Units   mg = milligram

Niacin can be purchased in tablet form at drug stores, and is a common ingredient in poultry vitamin mixes. Adding 5 to 7.5 pounds of livestock grade brewer's yeast per 100 pounds of chicken feed (or two to three cups of yeast per ten pounds of feed) will also prevent a niacin deficiency in ducklings.

Green Feed

The daily feeding of leafy greens to ducklings fortifies their diets with essential vitamins and minerals, reduces feed consumption and lowers the possibility of cannibalism. Tender young grass (before it joints), lettuce, chard, endive, watercress and dandelion leaves are excellent green feeds.

Ducklings will eat their salad *only* if it is tender and fresh. When greens are placed on the floor of a brooder, they soon wilt and are trampled and soiled. By putting the chopped feed in the ducklings' water trough, it remains succulent and clean, and the ducklings spend many contented hours dabbling for the bits of greenery.

Grit

Coarse sand or small granite grit should be kept before ducklings at all times. Grit aids the gizzard in grinding, helping birds to get the most out of their feed.

Feeders

For the first day or two, feed should be placed in containers where the ducklings cannot help but find it. Jar lids, shallow cans and egg flats are excellent for this purpose. Once the ducklings have located their feed and are eating well, they should be fed in trough feeders to reduce wastage.

Feeders can be purchased or constructed at home out of scrap materials. Homemade troughs should be designed so they do not tip over as ducklings jockey for eating space. To keep birds out of the feeder, a spinner can be attached across the top.

Sufficient feeder space needs to be provided to insure that each duckling receives its share. When limiting feedings to one or two daily, supply adequate trough space so that each bird can eat without having to struggle for its portion of the meal. By filling troughs no more than *half full*, you can significantly reduce the amount of feed that is wasted.

Creep Feeding

When ducklings are brooded by hens and run with the flock, it is sometimes desirable to feed the young birds separately from the adult ducks. This is done

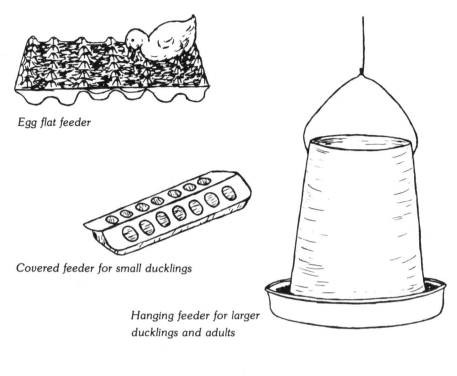

Egg flat feeder

Covered feeder for small ducklings

Hanging feeder for larger ducklings and adults

Homemade feeder with spinner and section for grit

by devising a creep feeder, where ducklings can eat without competing with grown birds.

The basic component of the creep feeder is a doorway large enough for only ducklings to pass through and gain access to the feed trough. The doorway panel can be placed across a corner of the duck yard or in the entrance of a shed.

Mature ducks can squeeze through smaller holes than is generally realized, and I have found that there is a tendency to make the slots too large. Dimensions for the portals will vary according to the breed raised, and you will probably need to do a little experimenting to find what size works best for your ducks. However, as a general rule, a space four inches square is satisfactory when breeds weighing seven to nine pounds are kept. For smaller breeds, such as Mallards, the passageway needs to be approximately three inches square to keep the old birds from entering and chowing with the youngsters.

MANAGING "GREEN" DUCKLINGS

"Green" ducklings are the equivalent of broilers in chickens. These birds are managed to produce the quickest possible growth in the shortest period of time on the least quantity of feed. The breed best suited for this practice is the Pekin. With good management, they are capable of weighing seven pounds at seven weeks of age on approximately twenty pounds of feed.

To stimulate fast growth, "green" ducklings must have limited exercise, a continuous supply of high-energy, concentrated feed, and twenty-four hours of light daily. They are ready to be butchered as soon as their primary wing feathers are developed. If held beyond seven or eight weeks, their feed conversion decreases *rapidly* (see Table 14) and they'll commence to molt, making it hard to pick them until they are fourteen to sixteen weeks of age.

PRODUCING LEAN DUCKLINGS
FOR MEAT

The major drawback of raising ducks for meat is the high fat content of their carcasses. While many people enjoy the succulency of quick-grown duckling meat, recent medical research strongly indicates that many of us need to reduce the amount of fat in our diets. There are several management practices that can be used to produce leaner ducklings.

One method is to use a high protein grower ration that contains an energy:protein ratio of 65:1 or less (see Appendix A, page 135). While ducklings fed such rations do not gain *weight* as rapidly as birds given feed with a wider energy:protein ratio (88:1 is normally recommended), the actual amount of *meat* will be almost identical.

A second method, which is more effective and practical for the small flock owner, is to raise a breed other than Aylesbury or Pekin; Muscovies, Mallards, Indian Runners and Campbells are generally considered the leanest. Then, rather than pushing ducklings for the fastest possible growth, they should be allowed to forage for a portion of their food. The birds should be given just enough feed to keep them healthy and growing well. When butchered at twelve to twenty weeks of age, the fat content of birds raised in this manner is similar to that of wild ducks (see Table 6, page 11). In a trial involv-

TABLE 14
TYPICAL GROWTH RATE, FEED CONSUMPTION AND FEED CONVERSION
OF PEKIN DUCKLINGS RAISED IN SMALL FLOCKS

Treatment	Age (weeks)	Live Weight (pounds)	Feed per Bird (pounds)	Feed per lb. of Bird (pounds)
Fed free choice on a 20%	5	5.07	10.85	2.14
protein, 1400 kcal/lb ration	6	6.10	14.52	2.38
from 0 to 2 weeks; then	7	6.85	18.43	2.69
fed 16% protein, 1400	8	7.43	22.36	3.01
kcal/lb ration from 3 to 12	9	7.87	26.52	3.37
weeks. Exposed to 24	10	8.36	31.43	3.76
hours of light daily and	11	8.58	36.47	4.25
raised in confinement.	12	8.43	40.55	4.81

ing thirty Rouen drakes from a production-bred strain, we found that when this method was employed, even these large meat birds produced carcasses that were no more fatty than roasting chickens.

SEXING DUCKLINGS

Over the years, I have been told many secret methods for sexing day-old ducklings. However, when tested, many of these techniques have proven to be only 50 percent accurate at best. To my knowledge, the following methods are the most trustworthy procedures for sexing live ducklings.

Vent Sexing

The *only* sure way to sex ducklings of all breeds before they are six to eight weeks old is by examining the cloaca. While it is much easier to vent sex waterfowl than land fowl, this procedure still requires practice, an understanding of the bird's physiology, and care to avoid permanent injury to the bird. Too often, persons attempt to sex ducklings without first acquiring the needed skills, injuring the birds or wrongly identifying the gender.

At best, a written account on sexing ducklings is a poor substitute for a live demonstration by an experienced sexer. I *highly recommend* that you have a knowledgeable waterfowl breeder show you how to vent sex ducks before trying it yourself.

While ducks of all ages can be sexed by this method, I suggest that novices practice on birds that are two to three weeks old. At this age, ducklings are easily held and their sex organs are large enough to be readily identified. As birds reach maturity, the sphincter muscles which surround the vent become stronger, making it more difficult to expose the cloaca. Extremely young ducklings are usually harder for the beginner to hold for sexing, and the sex organs are tough to identify until a person knows what to look for. There are three points to keep in mind when vent sexing ducklings:

1. Young ducklings are *extremely* tender and clumsy fingers can kill the bird or permanently injure the sex organs.

2. A bird has not been sexed *until* the cloaca has been exposed. People often assume that if they can't find a penis after applying a little pressure to the sides of the vent, the bird is a hen. However, a duckling cannot be identified as a hen until the cloaca has been inverted and no penis is evident.

3. The sex organs of ducklings are tiny, so it is *essential* that birds be sexed outside on a sunny day or under a bright light.

After studying the accompanying illustrations, you should be able to sex ducklings if the following steps are carefully implemented. There are several methods for holding birds for sexing, but I'll describe only the way I find most comfortable.

Step 1. Hold the bird upside down with its head pointed toward you. If the duck is large, its neck and head can be held between your legs.

Step 2. Use the middle and/or index finger of your right hand to bend the tail to a position where it is almost touching the bird's back.

Step 3. Push the fuzz or feathers surrounding the vent out of the way so you can see what you are doing.

Step 4. Place your thumbs on either side of the vent and apply pressure down and slightly out.

Step 5. Use your index finger of the right hand to apply pressure down and out on the backside of the vent. This inverts the cloaca and exposes the sex organs.

If it's a drake. If the duckling you are sexing is a male, a corkscrew-shaped penis will pop up near the center of the cloaca. (You must look carefully, since sometimes only the tip of the penis will be visible.) In drakes only a few days old, the penis is extremely small and almost transparent. By the time

*Vent Sexing
the Small Duckling*

1. Vent sex young ducklings under good lighting. Hold the bird upside down with its head pointed toward you and double its tail back with an index finger.

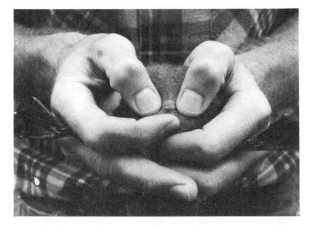

2. With the legs out of the way, position your thumbs on opposite edges of vent and place free index finger just behind the vent. To invert the cloaca, simultaneously apply pressure down and out with thumbs and index finger.

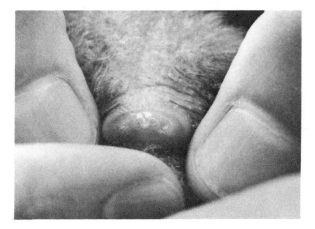

3. Almost transparent, the penis of a week-old drake is visible in the center of the inverted cloaca. A duckling cannot be identified as a hen until the cloaca is inverted as shown and no penis is present.

Vent Sexing the Older Bird

1. A convenient method for restraining large ducks for sexing is to turn the bird upside down and hold its neck between your legs.

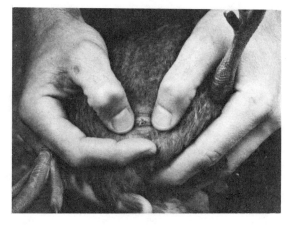

2. After bending the tail back, push aside the feathers surrounding the vent, position your thumbs on the sides and your index finger on the back of the vent, and invert the cloaca by applying pressure down and out.

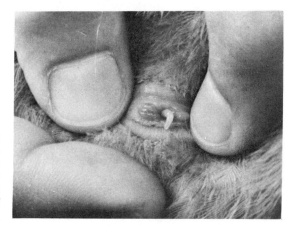

3. The cloaca and penis of an eight-week-old drake. If the cloaca is inverted sufficiently to expose the two dimples visible in this photo and still no penis is evident, it can be assumed that the bird is a hen.

males are a couple weeks of age, their sex organs are easier to see because of their larger size and deeper white or yellowish color.

If it's a hen. Should you have a hen, no penis will be visible when the cloaca is inverted. In hens that are more than several weeks old, the female genital eminence is often visible as a small, dark (usually gray) protuberance that resembles an undeveloped penis.

If you have difficulty. The most common problem encountered by unexperienced sexers is getting the cloaca inverted. If this is your problem, make sure that:

1. You have the bird's tail bent back double as in the illustration.
2. You are applying pressure down and out simultaneously with both your thumbs and the right hand index finger.

As with most skills, your speed and accuracy in sexing will improve with practice. If you get discouraged on your initial attempts, wait a few days until you have regained your confidence and then try again.

Sexing by Voice

By five to eight weeks of age, hens of most breeds can be distinguished by their voices. To sex birds by this method, catch each duckling individually, and as it protests its predicament, listen carefully for the hen's distinctive quack.

Sexing by Bill Color

It is possible to sex purebred ducklings of some varieties by their bill color as early as six weeks of age. In gray varieties such as Mallards and Rouens, the bill of a drake will normally turn dull green while that of a hen becomes dark brown and orange. The hen in most other colored varieties has a darker bill than the drake. This method cannot be used reliably on some crossbred ducklings since variation in bill color can be a sign of genetic differences rather than gender.

Sexing by Plumage

Drakes do not acquire their adult nuptial plumage until they are four to five months old. However, in many colored varieties, there are sufficient variations in the juvenile feathering of drakes and hens that it is possible to differentiate the sexes of ducklings that are five to eight weeks old. In gray varieties such as Mallards and Rouens, the feathers of the crown of the head and back often are darker and have less brown penciling in drakes than in hens.

MARKING DUCKLINGS

It is sometimes desirable to mark certain birds so they can be easily distinguished from the rest of the flock. Several methods can be utilized, including the use of leg or wing bands, and the notching and perforating of the webbing between the duckling's toes. I prefer to use bands since they do not permanently disfigure the duck.

Bands and toe punches are available from poultry equipment suppliers and feed stores. When leg bands are applied, care must be taken to use rings that are large enough to permit *free* circulation of blood to the feet.

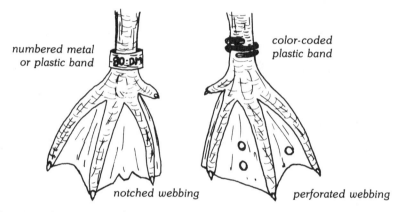

numbered metal
or plastic band

color-coded
plastic band

notched webbing   perforated webbing

Marking birds for easy identification

REVIVING WATERLOGGED DUCKLINGS

If you raise ducks, there's a good chance that you'll have at least one case of ducklings becoming soaked. In this condition, the little ones chill and quickly lose mobility. However, they can often be revived if you take *immediate* action, even when there is scarcely any movement left in their bodies.

The first step is to rub down the waterlogged birds with a soft cloth. Then put them in a warm (90° to 95°F.) place such as under a brooder or heat lamp, being careful not to *overheat* them.

Once they are up and around, provide feed and lukewarm drinking water. Keep the ducklings isolated until you are sure they have regained sufficient strength to be returned to their broodmates.

BROODING DUCKLINGS WITH OTHER BIRDS

Many of us who have duck flocks also raise other poultry. Consequently sometimes we're brooding ducklings and other birds at the same time. To save space and equipment, it is tempting to brood the various species together.

However, the young of ducks, geese, quail, guineas, chickens and turkeys have unique habits, temperaments, growth rates and management requirements that can cause serious problems when birds are raised in mixed groups. While it is not impossible to brood ducklings with other young poultry—goslings and ducklings get along fairly well—many problems are avoided and each grows better if they are brooded separately.

CHAPTER 9

Managing Adult Ducks

Adult ducks do not require a lot of time-consuming care or specialized facilities. By using common sense and being observant, novices can manage ducks successfully.

BASIC GUIDELINES

The most common error made with ducks is trying to raise them like chickens. For good results, every type of bird must be managed somewhat differently.

Protect your birds from predators. Ducks raised in small flocks seldom die from disease or exposure to severe weather, but quite a number are lost to predators. Every night, they should be penned up in a building or a securely fenced yard.

Supply a balanced diet and avoid medicated feeds. Following close behind predators, improper nutrition is the second leading cause of problems encountered by duck keepers.

Provide a steady supply of drinking water. Ducks do not thrive if they are frequently left without water.

Furnish suitable living conditions. Keeping ducks locked up in yards covered with deep mud and stagnant waterholes is an invitation for trouble.

Do not disturb them more than necessary. Waterfowl thrive on tranquility.

When catching, do not grab or carry them by their legs. The legs and feet of ducks are easily hurt.

HOUSING

Ducks require minimal housing. Unlike chickens, they prefer to stay outside day and night in most weather. In mild climates it is possible to raise ducks without artificial shelters. A windbreak made of straw bales will provide sufficient protection to keep ducks comfortable in regions where temperatures occasionally dip to 0°F. A more substantial shelter is needed in areas where extremely low temperatures are common.

Because waterfowl normally perch on water or the ground at night, the main reason for housing ducks in mild weather is to protect them from night-wandering predators (see Appendix C, page 147). A tight fence at least four feet high enclosing the yard is enough to stop many predators. However, in areas where thieves such as weasels, mink, raccoons, owls and wild or domestic cats are common, it is much safer to close ducks in a varmint-proof building or pen at nightfall.

If you do not have an empty building that can be converted into a duck-house, an inexpensive shed-like structure can be constructed. In the illustration, a practical duckhouse is shown. This type of shelter is portable and can be used for either ducklings or adult birds.

Since ducks are short, there is no need to have the walls of the duckhouse more than two feet high at the lowest point, except for your convenience in

Windbreak made of straw or hay bales

Practical duckhouse with attached nests

gathering eggs and cleaning out old litter. By having the nests attached along
the outside and outfitted with a hinged top, eggs can be gathered without entering the shed. Three solid walls and a wire front are recommended except in
regions with severe winters, where it is better to have a closed front. Good
ventilation is essential though, even in cold climates, since ducks fare poorly if
they are forced to stay in stuffy, damp quarters.

Dirt or sand floors are preferred to cement or wood floors in duck buildings.
To insure good drainage, place the shed on a slope or build the floor up four
to eight inches above ground level with dirt, sand or bedding. If the duckhouse is located in a low spot, water will accumulate, making it impossible to
keep the litter dry.

When ducks are housed only at night, allow at least 2.5 to 4 square feet of
floor space per bird. If you anticipate keeping your ducks inside continuously
during severe weather, a minimum of six to eight square feet of floor space
should be provided.

Cold Weather Housing

Ducks are well-prepared to remain comfortable in freezing weather with
their thick garb of feathers and down. Nonetheless, feed consumption can be
reduced and egg production increased if ducks are protected from the sever-
ity of bitter northern winters, particularly at night when the birds are inactive.

Insulated quarters. Insulating a small duckhouse is inexpensive and can
mean the difference between few or many eggs during the long winter
months. If the duckhouse has double wall construction, air spaces can be
filled with sawdust, chopped straw or a commercially prepared material.

Another satisfactory way to insulate shelters is to stack straw bales against the outside walls from ground level to the roof.

Deep litter. A thick layer of bedding in the duckhouse keeps ducks off cold floors and reduces the penetration of cold from the ground during periods of low temperatures. The deep litter system actually goes one step further by producing heat through decomposition of the litter and manure. This system has been tested in cold climates and has been found to increase winter egg production by up to 20 percent. It also eliminates the need to clean out duckhouses more than twice yearly, and produces excellent organic matter for the home garden.

To employ the deep litter method, follow these three steps: (1) cover the duckhouse floor with four to six inches of dry bedding; (2) periodically stir the bedding and add fresh material to keep the floor in good condition; and (3) clean out the old litter each spring. Because the litter will accumulate to a depth of eighteen to thirty inches, the walls of duckhouses where this system is used need to be approximately four feet high.

The Duck Yard

Most serious duck raisers have a fenced yard where birds are locked in at nighttime, or continuously if space is limited. The ideal yard has a minimum of ten to twenty-five square feet of ground space per duck, natural or artificial shade, a slope that provides good drainage, and a surface that is free of stagnant waterholes and deep mud.

If the soil on the land where you live has poor drainage, the duck yard should be covered with pea gravel, sand, straw, wood shavings or leaves, with the center of the yard built up higher than the outside edges. When choosing a location for your duck yard, keep in mind that a dense population of birds can kill young trees due to the high nitrogen content of duck manure.

Keeping Ducks on Wire Floors

Adult ducks can be kept in houses or hutches with wire-covered floors or in elevated, all-wire cages. The main advantage of this system is that birds can be maintained in extremely limited space without problems of sloppy litter or muddy yards. Some disadvantages are that feed consumption is higher since the birds cannot forage, foot problems can develop (particularly among the larger breeds), and life is less enjoyable for the ducks.

To prevent lameness, wire mesh used on floors should not have openings larger than 1 inch by ½ inch. However, if finer mesh is used, manure may not pass through, causing buildup problems. A minimum of two square feet of floor space per bird should be allowed, with double or triple this amount being recommended when feasible.

Bedding

Ducks can withstand wet weather, but they should *never* be forced to stay in muddy, filthy yards or buildings. The floors of buildings should be covered with several inches of absorbent litter. In muddy or snow-covered yards, mounds of bedding four to six inches thick should be provided to give ample space where all ducks can roost comfortably. Add new layers of clean bedding as needed.

Shade

Ducks must have access to shade in hot weather. They are a cool-weather bird and suffer if forced to remain in the sun at temperatures above 70°F. When ducks are confined to a yard lacking natural shade, a simple shelter should be provided which will supply adequate shade for all birds residing there. Feed troughs can also be located under this cover to protect the feed from exposure to the elements.

A simple shelter provides shade and protects feed from sun and rain.

Nests

Clean eggs hatch better and retain their freshness for eating longer than eggs that must be washed. Having adequate nests for your hens will help produce clean eggs and reduce the chances of eggs being broken. So that hens are familiar with them, install nests two weeks before you expect the first eggs. One nest for every four or five hens is sufficient.

For medium to large hens, nests approximately a foot square and twelve inches high are recommended. Duck nests are placed on the ground and do

Indoor nests

Outdoor nests

not need solid bottoms. If nests are inside buildings, they do not require a top. Covering the nest bottoms with burlap aids in the production of clean eggs. Keeping nests well-furnished with clean, dry nesting material such as sawdust or straw will encourage hens to use them and result in fewer broken or soiled eggs.

Portable Panels

Lightweight panels constructed from 1″ × 2″ or 1″× 4″ lumber and chicken netting frequently come in handy. They can be used to divide pens, for isolating hens that are setting or have young, as a catch pen in a large yard, and for small pens that can be moved around to utilize grass in different areas of the home place. To retain most breeds of ducks, the panels only need to be two or three feet high. Panels that will be used to impen day-old ducklings should be covered with 1″ × 1″ chicken netting.

SWIMMING WATER

Ducks enjoy swimming, and add charm and interest to lakes, ponds and brooks. Where natural bodies of water do not exist, small earthen or cement ponds can be constructed. It is advantageous to design them for easy drainage and cleaning. A four-inch-deep layer of sand or gravel around the

White Calls taking it easy on a small cement pond.

perimeter of a pond is helpful in keeping ducks from tracking in mud and drilling with their bills.

Contrary to popular belief, ducks can be raised successfully without water for swimming. In fact, there are some advantages in having only drinking water. If a small pond is overcrowded, sanitation problems often arise. Ducks must *not* be allowed to swim in or drink filthy, stagnant water.

DRINKING WATER

Ducks must have a *constant* supply of reasonably clean drinking water. So the birds can wash out their nostrils, waterers should be designed to provide water that is at least three inches deep. Buckets, dishpans and hot water tanks cut in half vertically make satisfactory drinking containers for adult ducks. For a pair or two of birds, a one-gallon tin can will suffice.

Because ducks frequently wash their bills, drinking water would need to be changed many times daily to be kept clear. While such repeated changing is not necessary, the water should be replaced several times weekly and must never be allowed to become putrid.

To keep unhealthful mudholes from developing around watering areas, water fountains should be placed on wire-covered platforms. For additional protection, a pit twelve to twenty-four inches deep can be made underneath the platform and filled with gravel.

During cold weather when there is soft snow, ducks can sustain themselves

*Water trough on
wire-covered platform*

for a time by eating the icy crystals. However, it is preferable to give them warm drinking water twice daily when the temperature is cold enough to freeze all available liquid.

RAISING DUCKS ON SALT WATER

Ducks can be raised successfully in marine areas. Most ocean bays and inlets are teeming with an abundant supply of plant and animal life that ducks relish. Because domestic ducks have a lower tolerance for salt than do wild sea ducks, sweet drinking water should be supplied at all times. Birds that are raised for meat on marine waterways need to be confined in a pen or yard and fed a grain-based diet for two to four weeks prior to butchering to avoid fishy-flavored meat.

NUTRITION

Feed is the single most expensive item in raising ducks, normally representing 60 to 80 percent of the total cost. Finding ways to save on feed expenditures will significantly decrease the cost of your duck project.

For top egg production, ducks must be fed an adequate amount of concentrated feed having 16 percent crude protein. Three weeks before the first eggs are expected and throughout the laying season, a laying ration should be fed free choice or twice daily. (See Table 15 for recommended daily feed allowances.) Duck hens lay the best when they are in a semi-fat condition, but excessive fatness is harmful and must be avoided. A sudden change in the diet of hens that are laying normally results in a sharp decline in egg production, throwing hens into a premature molt from which it will take six to ten weeks to recover.

TABLE 15
SUGGESTED FEEDING SCHEDULE FOR ADULT DUCKS

Size of Duck	Holding Period When Birds Are Not Producing Lbs. of 12-14% Protein Feed per Bird Daily	3 Weeks Prior to and During Laying Season Lbs. of 16% Protein Feed per Bird Daily	2 Weeks Prior to Butchering Mature Ducks Lbs. of High-Energy Feed per Bird Daily
2-3 lbs.	0.15-0.25	0.20-0.30	Free choice
4-5 lbs.	0.20-0.30	0.30-0.40	Free choice
6-7 lbs.	0.25-0.35	0.40-0.50	Free choice
8-9 lbs.	0.30-0.45	0.45-0.60	Free choice
Muscovy	0.30-0.40	0.40-0.60	Free choice

The quantities of feed that you need to supply ducks are highly dependent on the availability of natural foods, the climatical conditions and the quality of feed (e.g., birds require larger amounts of high fiber foods than low fiber foods to meet their energy requirements).

While not in production, mature ducks can be given a maintenance ration containing 13 to 14 percent crude protein, and fed just enough to keep them in good condition. The quantity of feed required by birds is highly affected by the weather. During cold periods, ducks must eat considerably more feed than when temperatures are moderate.

Fish by-products such as fish meal are excellent sources of high-quality proteins, and are used in many commercially prepared poultry feeds. These ingredients are excellent for immature birds and adult breeders that produce hatching eggs, but laying hens that supply eating eggs must be given feeds that are low in (less than 4 percent), or do not contain, fish products.

Feeding Programs

In formulating a feed program for your adult duck flock, try to find ways to obtain a high degree of productivity while utilizing food resources that normally go unharvested.

Natural. The least expensive way to feed ducks is to make them forage for their food. This option is limited to situations where wild and natural foods are in abundant supply and when top egg production is not important. It must be remembered that the quantity of natural feed fluctuates widely during the various seasons of the year. While there may be periods when ducks can find most and possibly all of their own feed, there will probably be times when most of their food will need to be supplied in the form of grain and mixed rations. Ducks must *never* be allowed to deteriorate and become thin due to lack of feed. Negligence in this area can permanently damage their productivity.

Grains. Whole grains, by *themselves,* are not a complete poultry feed. However, if ducks are given a protein concentrate (see Appendix A, page 135) or can forage in waterways or pastures, a supplement of grain will often satisfy their dietary needs, particularly when the birds are not in production.

Ducks are not as fond of oats and barley as they are of other grains. But these two grains are good waterfowl feeds and the birds will learn to eat them if forced to do so.

Corn is a high energy grain and an excellent feed for cold weather and for fattening poultry—if you desire fat meat. During hot weather, the diet of ducks should not consist of more than 60 to 80 percent corn. If waterfowl are fed too much corn during exceptionally hot weather, egg production drops and health problems can arise.

Home mixed. It may be practical to mix your own feed if the various ingredients are available at a reasonable price. Mixing a home ration for ducks is simpler than for chickens. All ingredients for chicken feeds need to be finely ground; otherwise the birds will pick out the grains and leave the finer particles. This problem is usually not encountered with ducks since they tend to scoop up their feed rather than picking it up one kernel at a time. (See pages 77-78 for methods of mixing feed.)

The formulas in Table 16 are examples of the type of feeds that can be mixed at home for small flocks of ducks. If a commercially prepared vitamin:mineral premix is used, the rations in Table 17 can be prepared at home as well. The suggested ingredients can be substituted with similar products that are more readily available in your area. (See Appendix A, page 135 for instructions on how to formulate rations.)

Commercial. The simplest way to provide ducks with a balanced diet and stimulate top production is to purchase commercially mixed concentrated feed. These rations are formulated to insure a proper balance of carbohydrates, fats, proteins, vitamins, minerals and fiber.

In many localities, feeds manufactured specifically for breeding and laying ducks are not available. Nonmedicated turkey and chicken feeds can be used in place of duck mixtures. (Chicken rations usually need to be supplemented

TABLE 16
HOME-MIXED RATIONS FOR ADULT DUCKS

Ingredient*	No. 12 Holding pounds	No. 13 Holding pounds	No. 14 Layer pounds	No. 15 Layer pounds
Whole milo or yellow corn	82.00	—	60.00	24.00
Whole soft wheat	—	86.00	9.00	48.00
Soybean meal (50% protein)	8.00	4.00	7.00	4.50
Meat and bone meal (50% protein)	—	—	4.00	4.00
Alfalfa meal (17.5% protein)	4.00	4.00	4.00	4.00
Dried skim milk	—	—	2.00	2.00
Brewer's dried yeast	5.00	5.00	7.00	7.00
Oyster shell	0.25	0.50	6.40	5.90
Dicalcium phosphate (18.5% protein)	0.50	0.25	0.30	0.35
Iodized salt	0.25	0.25	0.30	0.25
Cod liver oil**				
Totals (lbs.)	100.00	100.00	100.00	100.00

*The addition of 3 to 5 pounds of livestock grade molasses to each 100 pounds of mixed feed reduces waste.
**If birds do not receive direct sunlight, which enables them to synthesize vitamin D, sufficient cod liver oil must be added to these rations to provide 400 International Chick Units (ICU) of vitamin D_3 per pound of feed.
Note: These rations are best suited for situations when birds have access to pasture.

with additional niacin. See page 78.) Ducks waste less feed if they are fed *pellets* rather than crumbles or mashes.

If you are raising a large number of ducks, it may be economical to have a local feed mill custom-mix your duck feed. In Table 17, feed formulas that we have used successfully for a number of years are given.

Because wheat is our most abundant and cheapest grain here in the Northwest, we use it as the base for our rations. In your locality, other grains may be lower priced and they should be used. Your grain dealer can help you formulate a feed that best utilizes available resources.

Growing Duck Feed

It is possible to raise a good portion of the feed required by a flock of half a dozen ducks on a 25′ × 25′ plot of land. Field corn yields large quantities of ears that are easily harvested and shelled by hand or broken in two and thrown to the ducks to shell for themselves. Grains such as wheat and rye provide fall and spring grazing and, when the seeds are mature, the ducks can be allowed into the patch for a short time daily to harvest their own feed. Grain sorghums such as kafir and milo produce large seed heads that can be hand harvested and stored whole.

A large assortment of natural crops is available for planting in and around bodies of water. Some of the favorites of ducks include wild celery, wild rice,

TABLE 17
COMPLETE RATIONS FOR ADULT DUCKS (PELLETED)

Ingredient	No. 16 Corn Base Holding lbs/ton	No. 17 Wheat Base Holding lbs/ton	No. 18 Milo Base Holding lbs/ton	No. 19 Corn Base Breeder lbs/ton	No. 20 Wheat Base Breeder lbs/ton	No. 21 Milo Base Breeder lbs/ton
Ground yellow corn	1612	—	—	1415	—	—
Ground soft wheat	—	1667	—	—	1460	—
Ground milo (grain sorghum)	—	—	1630	—	—	1421
Soybean meal solv. (50% protein)	265	190	255	302	236	295
Meat and bone meal (50% protein)	—	—	—	80	80	80
Alfalfa meal (17.5% protein)	40	40	40	40	40	40
Stabilized animal fat	—	30	—	—	20	—
Soybean oil	—	5	—	—	15	—
DL-Methionine (98%)	—	1.25	1.25	1.00	1.50	2.00
L-Lysine (50%)	—	1	—	—	—	—
Dicalcium phosphate (18.5% P)	28	27	27	12	10	12
Limestone flour (38% Ca)	30.00	13.75	21.75	125.00	112.50	125.00
Iodized salt	5	5	5	5	5	5
Vitamin:mineral premix	20	20	20	20	20	20
Totals (lbs.)	2000	2000	2000	2000	2000	2000

Vitamin:Mineral Premix

Vit. A (millions of IU/ton)	5.0	6.0	6.0	8.0	9.0	9.0
Vit. D_3 (millions of ICU/ton)	0.8	0.8	0.8	1.0	1.0	1.0
Vit. E (thousands of IU/ton)	—	10.0	10.0	10.0	20.0	20.0
Vit. K (gm/ton)	1.0	1.0	1.0	2.0	2.0	2.0
Riboflavin (gm/ton)	3.0	3.0	3.0	6.0	6.0	6.0
Vit. B_{12} (mg/ton)	4.0	4.0	4.0	8.0	8.0	8.0
Niacin (gm/ton)	30.0	30.0	30.0	50.0	50.0	50.0
d-Calcium pantothenate (gm/ton)	2.0	1.0	1.0	6.0	3.0	3.0
Choline chloride (gm/ton)	100.0	—	100.0	300.0	—	300.0
Folic acid (gm/ton)	—	—	—	0.5	0.5	0.5
Manganese sulfate (oz/ton)	9.6	9.6	9.6	9.6	9.6	9.6
Zinc oxide (80% Zn, oz/ton)	3.2	3.0	3.2	3.2	3.0	3.2
Ground grain to make 20 lbs.	+	+	+	+	+	+
Totals (lbs.)	20.0	20.0	20.0	20.0	20.0	20.0

Calculated Analysis

Crude protein (%)	14.2	13.6	14.1	16.3	15.8	16.2
Lysine (%)	0.63	0.60	0.63	0.77	0.72	0.75
Methionine (%)	0.26	0.26	0.26	0.33	0.33	0.32
Methionine + cystine (%)	0.50	0.53	0.49	0.58	0.58	0.57
Metabolizable energy (kcal/lb.)	1419	1359	1405	1322	1278	1306
Calorie:protein ratio	100	100	100	81	81	81
Crude fat (%)	3.3	3.4	2.5	3.2	3.6	2.6
Crude fiber (%)	2.6	2.8	2.7	2.6	2.7	2.7
Calcium (%)	0.95	0.64	0.77	2.99	2.75	3.00
Available phosphorus (%)	0.35	0.36	0.35	0.40	0.39	0.41
Vit. A (IU/ton)	5240	4678	4678	6610	6178	6178
Vit. D_3 (ICU/ton)	400	400	400	500	500	500
Riboflavin, (mg/lb)	2.17	2.21	2.20	3.73	3.76	3.76
Total niacin (mg/lb)	25	38	32	35	46	41

KEY: IU = International Units   gm = gram   kcal = kilocalories
ICU = International Chick Units   mg = milligram

wild millet, small bulrush, smartweed and chufa tubers. Addresses of nurseries specializing in natural game bird crops can be found in hunting and outdoor magazines. Ducks also enjoy and benefit from vegetable and fruit produce that cannot be used in the kitchen.

Pasture

While ducks can be raised in barren yards or pens, they enjoy succulent vegetation when it is available. Good quality forage lowers their feed consumption by approximately 10 percent and lessens the possibility of vitamin deficiencies. Ducks cannot eat mature pasture (except the seeds), so it should be mowed occasionally to encourage new growth.

Orchards are an excellent location for duck pastures. The vegetation smothers out noxious weeds and provides a protective and attractive ground cover. Ducks significantly reduce the numbers of harmful insects in orchards and clean up diseased and windfall fruits.

There are a variety of grasses and legumes that make good permanent pasture for ducks. The main requirements are that the plant thrives in your region and that it produces succulent forage. A mixture of Perennial Rye grass and one of the clovers such as New Zealand White, Lodino or White Dutch will provide good grazing in many localities. Your local agriculture extension service specialist can give advice on what varieties do well in your climate.

Feeding Insects to Ducks

Insects are a bountiful source of high-quality protein. While ducks of all ages are accomplished bug hunters, their consumption of winged insects can be increased by burning a low-wattage bulb twelve to eighteen inches above ground level in the duck yard at night. As the insects swarm around the light, the birds have hours of leisurely dining.

Utilizing Surplus Eggs and Milk

Eggs and milk are excellent sources of nutrition for poultry. Being "complete" foods, they are particularly valuable in home-mixed rations to help insure that birds are receiving a balanced diet. In general, when egg and/or milk are added to the diet of ducks, they grow better, have glossier plumage and lay more eggs.

The only practical way to feed liquid milk to ducks is to mix small quantities into a dry ration. When milk is given in pans or buckets, ducks play in it, covering themselves and the surrounding area with the sticky stuff. Under these conditions, food poisoning and eye infections often develop.

Eggs should *always* be hard-boiled before they are given to poultry. Feeding raw eggs can result in a biotin deficiency and often leads to birds eating their own eggs. To avoid spoilage, milk and eggs should not be mixed with the duck ration until feeding time.

Feeding Leftovers

Kitchen and garden refuse must be fed to producing birds in moderation. Leftovers normally are high in starch, fiber and liquid, and low in most other nutrients. Hens cannot lay well if their minimum requirements for protein, minerals and vitamins are not being ingested.

Grit and Calcium

To digest their feed to the best advantage, ducks need to have a continuous supply of granite grit, coarse sand or small gravel. Four weeks prior to, and throughout the laying season, hens need to have their diets supplemented with calcium if they are going to lay strong-shelled eggs. Most manufactured laying feeds contain the correct amount of calcium, but when grains or home mixes are used, a calcium-rich product such as egg shells, oyster shells or ground limestone should be fed free choice.

Feeders

Because adult ducks are often fed a limited quantity of feed, sufficient feeder space must be allowed so that *all* the birds can eat at one time without crowding. Otherwise, less aggressive and smaller ducks may be pushed aside and not get their full share.

Approximately six lineal inches of feeder space should be allowed per bird. Ducks can eat from both sides of most feeders, so a trough five feet long provides ten lineal feet of feeder space which is sufficient for a flock of twenty birds. V-shaped troughs, with spinners mounted along the top to keep birds out, are easily constructed and provide sanitary eating conditions.

Finishing Roasting Ducks

When ducks have been forced to forage for much of their feed during the growing period, they can be fed grain or finishing pellets free choice for two or three weeks prior to butchering. If they have ranged widely, it is advantageous to pen them in a restricted yard (allowing a minimum of twenty-five square feet per bird) during the finishing period. This time of heavy feeding and restricted exercise will encourage the ducks to gain weight rapidly, resulting in a larger carcass and succulent, tender meat.

THE LAYING FLOCK

Most duck hens lay before 8 in the morning. So that eggs are not lost in fields or bodies of water, it is a good practice to confine the laying flock to a yard or building at night.

When ducks are penned up at nighttime, they *must* have access to drinking water if feed is in front of them. Otherwise, they will suffer from thirst and may choke to death.

Unless eggs are going to be used for hatching, there is no need to have drakes. Hens lay better without frequent mating activity, and nonfertile eggs store longer than fertile ones.

Lighting Needs

When top egg production is desired during the short days of fall, winter and early spring months, duck hens, like chickens, *must* be exposed to artificial lighting. A minimum of fourteen hours of light daily is required to keep duck hens laying well. It is essential that the length of daylight *never decreases* while hens are producing, or the rate of lay will be severely diminished. Even a reduction of only fifteen to thirty minutes per twenty-four hour period for several days can negatively effect heavily producing hens.

By using an automatic time switch that turns lights on before daybreak and off after nightfall, hens can be exposed to fourteen to sixteen hours of light daily. Another method is to leave a light burning all night long. Since ducks become frightened quite easily at night and may wear themselves out by milling about, this system has the advantage of keeping them calm and quiet.

TABLE 18
TYPICAL EFFECTS OF MANAGEMENT ON EGG PRODUCTION

Treatment	Annual Egg Production per Hen		
	Domestic Mallards	Commercial Rouens	Khaki Campbells
Fed whole or cracked grains Given access to pasture & pond Exposed to natural day length	25-40	50-65	75-150
Fed 16% protein laying pellets Given access to pasture Exposed to natural day length	50-75	75-100	175-225
Fed 16% protein laying pellets Given access to pasture Exposed to 16 hours light daily	85-125	125-150	275-325

However, egg production is normally lowered slightly when hens are exposed to twenty-four hours of continuous light daily, in contrast to fourteen to sixteen hours.

To prevent premature egg production, the amount of light young hens are subjected to needs to be watched carefully. When pullets are exposed to excessive light or increasing day lengths between the ages of twelve to twenty-two weeks, they will begin to lay before their bodies are adequately mature. Early laying (before twenty weeks of age) can result in a shortened production life, smaller and fewer eggs, and greater possibility of complications such as prolapsed oviducts.

The intensity of light required to stimulate egg production is relatively low (approximately one foot-candle at ground level, which usually equals one bulb watt per four square feet of floor area). When ducks are confined to a building or shed at night, one forty- to sixty-watt incandescent bulb six feet above ground level will provide adequate illumination for each 150 to 250 square feet of floor space. In outside yards, one 100-watt bulb with a reflector per 400 square feet of ground space is recommended.

We use the following lighting schedule with good success on hens that are hatched in March through July, and I recommend it if you desire maximum efficiency and production.

Age	Hours of Light Daily
0-8 weeks	24 hours.
8-22 weeks	Natural day length.
22 weeks	Add 2 hours of artificial light to natural day length.
23 weeks	Add 15 minutes of light weekly until 15 or 16 hours of total light is reached. Maintain a constant level until hens stop laying or they are force molted.

Identifying Producing Hens

Shortly before production commences, and throughout the laying season, the abdomens of hens swell noticeably. Hens that are in production can be identified by their large, moist vents and wide-spread pubic bones. As the season progresses, the color of the bills of high-producing hens fades. This phenomenon is especially noticeable in ducks with orange bills.

Encouraging Hens to Use Nests

To encourage duck hens to lay where you want them to, place dummy eggs in nest boxes several weeks before the beginning of the laying season. While nest eggs can be purchased, I have found that turning them out on a lathe is an enjoyable rainy day project. You do not have to worry about making them exactly the correct shape and size. Painting the wooden eggs makes them last longer and easier to clean. Any color will be accepted by the hens,

TABLE 19
IDENTIFYING PROBLEMS IN THE LAYING FLOCK

Symptom	Common Cause
Thin or soft-shelled eggs	Usually a vitamin D_3 deficiency; also high temperatures, abnormal reproductive organs or calcium deficiency.
Eggs decrease in size	A gradual decrease in size is common as the laying season progresses. However, excessive environmental stresses and dietary deficiencies accentuate the problem.
Odd-shaped eggs	Temporary malfunction of the reproductive organs; in some cases, abnormal oviducts.
Pale yellow yolks	A diet lacking carotene, which is supplied by products such as yellow corn and green plants. Hens fed wheat-based rations without access to pasture or greens usually produce anemic-colored eggs.
Bright orange-red yolks	Diets high in corn and/or green feeds.
Blood or meat spots in egg interior	Internal hemorrhage. Eggs fit for eating.
Blood on shell exterior	Ruptured blood vessel at the cloaca opening. Frequently occurs when young hens begin laying.
The back of head and neck bare of feathers; in extreme cases, skin is lacerated and scabby	Too many drakes, which results in excessive mating activity. Drakes sometimes have favorite hens.
Hens go into a premature molt	A sudden change in diet or lighting; birds frequently left without water; onset of hot weather.
Eggs lost in bodies of water, pastures or hidden nests	Hens not locked in a pen or small yard at night, or are turned out before 8 or 9 A.M.

but white or light-colored shades have the advantage of being more visible to birds.

Production Life of Hens

With good management, small breed pullets begin laying at twenty to twenty-four weeks, while large breed pullets require twenty-four to thirty weeks. Duck hens lay the greatest number of eggs in their first year of production, but normally show only a minor decrease in productivity during the second and third year when eggs are larger in size. Good hens often produce some eggs until they are four to eight years old.

Breaking Up Broody Hens

When hens become broody, their egg production falls off or ceases. To induce broodies back into production, isolate them in a well-lit pen without nests or dark corners, and provide drinking water and feed. In these surroundings, hens normally lose their maternal desires in three to six days and can be returned to the flock. Because broodiness seems to be contagious, remove hens from the flock *promptly* when they show the first signs of wanting to set.

Force Molting

Each year ducks lose their feathers and replace them with new ones. Hens normally do not molt while they are laying since their bodies cannot support the strain of growing feathers and forming eggs at the same time. Under natural conditions, this arrangement works out fine; the hen lays in the spring, hatches and broods her young until they are able to fend for themselves, and then she molts, with her new garb being ready for the cold winter months.

When pullets of the egg breeds are managed for top egg production, they begin laying at twenty to twenty-four weeks of age which typically is in the fall of the same year they are hatched. With proper management, hens will lay throughout their first winter, the following spring and summer and into the fall months without molting. By this time their feathers are worn and will not provide ample protection from the low temperatures and wet weather of their second winter. Because it is feasible to keep duck hens for up to three years, it is a good practice to force molt them at the end of their first and second years of production.

The laying flock should be force molted when egg production is down—normally during the hot summer months—and when they will have sufficient time to replace their feathers before the cool, wet weather of fall. Because ducks require six to ten weeks to molt and begin laying again, they should be force molted in June or July in most localities.

Hens are thrown into a molt by sudden changes in their diet and environment. The following schedule is suggested as a simple method for force molting the home duck flock.

1st day: Discontinue artificial lighting and remove all water and feed.

2nd day: Provide drinking water but no feed.

4th day: Commence feeding again, but substitute the laying ration with .25 pounds of grain (preferably not corn) per bird per day. This feed allotment is for hens weighing four to five pounds.

8th day: Switch to a 17 to 18 percent protein grower ration to insure a healthy growth of new feathers. Supply .25 to .33 pound of feed per bird per day.

28th day: Gradually replace grower feed with laying ration, and supply a minimum of fourteen hours of light.

SELECTING BREEDERS

To maintain the productivity of the home flock, ducks that are to be used as breeders must be chosen carefully. In selecting breeding stock, only birds displaying the following characteristics should be retained: robust health, strong legs, good body size and freedom from deformities. If purebred ducks are raised, breed characteristics for typical size, shape, carriage, color and markings should also be considered.

The following deformities and weaknesses are highly inheritable and will occasionally show up in duck flocks. Birds with any of these faults *should not* be kept for reproduction.

Weak legs. While this problem can usually be traced to nutritional deficiencies, weak legs can also be inherited. Genetic leg weakness is often indicated by birds hobbling about with one or both legs twisted slightly outward or birds whose legs give out quickly after walking short distances.

Crooked toes. One or more toes are bent sharply at an unnatural angle, making this disfigurement easily identified.

Slipped wing. Probably the most common deformity seen in ducks, slipped wings are identified by wingtips that stick out from the sides rather than folding smoothly against the body. (See page 128.)

Wry tail. Rather than pointing straight back as is normal, wry tails are constantly cocked to one side.

Roach back. A deformed spinal column causes a humped and shortened back.

Kinked neck. This condition can be the result of an injury or from forcing tall ducks such as Indian Runners to remain in low boxes for extended periods of time (e.g., when they are shipped). Crested ducks exhibit necks with severe crooks just behind the head more than other breeds.

Crossed bill. When a duck has this defect, the upper mandible is usually bent to the side and not aligned with the lower mandible.

Scoop bill. Scoop-bills have an unnaturally deep, concave depression along the top line of the bill.

Blindness. Certain strains of poultry, particularly some of those that are highly inbred, show a high incidence of clouded or white pupils. Some birds will exhibit this fault at hatching time, while in others, it may not develop for several months or years.

CATCHING AND HOLDING DUCKS

Ducks *must be handled with care* since their legs and wings are injured easily. When catching, avoid running them on rough ground or where they will trip over feed troughs and other obstacles. It is advantageous to walk waterfowl into a small catching pen or V-shaped corner rather than having a wild chase around a large yard.

Never grab ducks by their legs or wings. Rather, grasp them securely but gently by the neck or with one hand over each wing to subdue them, and then slide one hand under the breast and secure the legs. When the bird is lifted from the ground, its weight should be resting on your forearm with its head pointed back between your body and arm, and its wings pinned against your side. When the wings and feet are held securely, there is little possibility that either you or the duck will be injured. Small and medium-sized ducks can be picked up and held with a thumb over each wing and the hands encircling the body.

Special precautions should be taken when handling Muscovies. They are surprisingly powerful and have long, sharp claws. When handling any type of bird, hold them away from your face to eliminate the possibility of injury to your eyes.

A safe and convenient method for catching ducks is to grasp the bird gently by the neck.

Small and medium-sized ducks such as this young White Campbell can be caught and held with a thumb over each wing and the hands encircling the body.

110

My father demonstrates two methods for holding ducks that are comfortable and safe for both the bird and the holder.

DEALING WITH AGGRESSIVE BIRDS

Ducks are not aggressive by nature. On rare occasions, however, a bird will develop a belligerent disposition and make a nuisance of itself by pecking and wing-slapping the legs of anyone who approaches too near. This problem is usually the result of ducks being teased and harrassed and can be avoided by teaching your birds to trust you.

Rehabilitating a rogue duck requires patience and an understanding of his nature. Many people aggravate the problem by kicking or striking back at pursuing birds. This action challenges the duck and stimulates it to fight harder. I have found that when a bird makes threatening advances, the best strategy is to ignore it and quietly go about your business, being careful not to make sudden moves. While some persons find this advice hard to follow, it usually works like magic. Without a stimulus to fight, the bird quickly loses interest and goes on its way.

If you happen to run into a hard-core rascal who refuses to mellow, you'll have to decide whether to put up with it or enjoy duck stew!

DISTINGUISHING THE SEX
OF MATURE DUCKS

The gender of adult ducks is easily recognized by secondary sex characteristics such as voice, feather formation, plumage color and body size. The voices of drakes resemble a hoarse whisper, in contrast to the quacking of hens. Except during the molt, drakes display several curled tail feathers, which hens lack. In colored varieties, drakes are brighter than hens, although for approximately two months each year, males undergo an eclipse molt during which time both sexes are similarly colored.

Muscovies are practically mute, although the hens can quack weakly. Drakes lack the curled tail feathers of true ducks, and the two sexes are similar in color. However, the sexes can be readily identified by the drakes' larger size and the hens' smaller facial skin patch.

CLIPPING WINGS

To keep flying ducks from wandering, it may be necessary to clip their wings. This is easily and painlessly accomplished by cutting the main flight feathers of one wing with tin shears or heavy-duty scissors. So that birds do not look unbalanced, the two outermost flights can be left intact. Once a year ducks molt their wing feathers and replace them with a new set. To keep them grounded, trim their wings annually.

You can ground flying ducks by clipping the feathers of one wing.

HANDLING MANURE

Fortunately, the handling of duck manure does not need to be a back-breaking chore. If you plan ahead, the birds will spread much of their droppings directly onto the land where it is needed.

In the fall of the year, after the garden crops have been harvested, a portion of, or the entire garden can be converted into the duck yard. A layer of bedding such as cornstalks, leaves or straw should cover the ground to a depth of at least three to four inches. This covering not only keeps the ducks out of mud, but it also protects the dirt from being packed down and provides an excellent environment for earthworms and soil-enriching microorganisms.

In the spring, the partially decomposed bedding and manure can be worked into the soil. In gardens where this method has been used, lush crops of vegetables and fruits have flourished with minimal use of commercial fertilizers.

CHAPTER 10

Butchering

One of the main reasons ducks are raised is for their excellent meat. While they can be taken to a custom dressing plant to be butchered, this service is often expensive and robs you of the satisfaction of preparing your own food.

Cleanliness throughout the butchering process is essential to curb contamination and spoilage of meat. All cutting utensils should be sharp—dull knives are a waste of time as well as unsafe.

WHEN TO BUTCHER

One of the most time-consuming parts of the butchering process is the removal of the feathers. To make this job as simple as possible, butcher ducks when they are in full-feather. If a duck is slaughtered when it is covered with pin feathers, a picking job that normally takes three to five minutes can develop into a frustrating, feather-pulling marathon.

Depending on the breed and management, ducklings are normally in full-feather at seven to twelve weeks of age, except Muscovies, which require fourteen to sixteen weeks to feather out. Within a short time—sometimes a matter of days—ducks go into a heavy molt, replacing their juvenile garb with adult plumage. If ducklings are not dressed before this molt commences, butchering is best delayed for six to ten weeks when their adult plumage will have been acquired. The flesh of ducks is not in prime condition while they are molting.

Prekilling Preparations

Ducks should be taken off feed four to six hours prior to killing, or the night before if they are going to be dressed early the following morning. To avoid excessive shrinkage, drinking water should be left in front of the birds until they are slaughtered.

Killing

For most of us, the least enjoyable task in raising ducks is killing them on butchering day. But for those of us who raise our own meat, it is a necessary chore.

There are several ways of dispatching ducks. The simplest and most impersonal method is the ax and chopping block. It is advantageous to have a device (such as two large nails driven into the block to form a V) on the chopping block to hold the bird's head securely in place. A sharp cutting edge on the ax is a must. As soon as the head is removed, the bird should be hung by its legs to promote thorough bleeding and to prevent it from becoming bruised or soiled from thrashing about on the ground.

A second method for killing is to suspend the live duck head-down with leg shackles or in a killing cone. Grasp the bill firmly with one hand; with the other stun the bird with a stout stick. With a sharp knife, cut the throat about an inch below the base of the bill, on the left side, severing the left jugular vein. The head provides a convenient handle if the duck is going to be scalded.

To avoid discolored meat in the dressed product, it is vital that ducks are bled *thoroughly* before they are processed further.

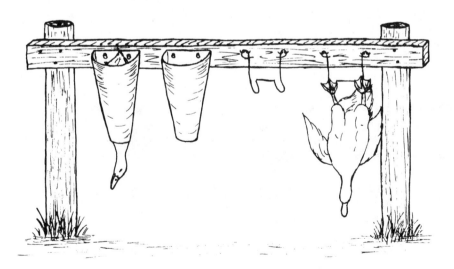

Killing cones and leg shackles

PICKING

The sooner a duck is picked after it is bled, the easier the feathers will come out.

Dry Picking

The best quality feathers for filler material and the most attractive carcasses are obtained when ducks are dry picked. Novices often find this method unbearably slow. Veterans, on the other hand, can dry pick a duck clean in three minutes or less. One secret for success is to extract feathers in the same direction they grow. Pulling feathers against the grain invariably results in torn skin. For a better grip, lightly dust the duck with resin, or periodically moisten your hands with water. The large plumes of the wings and tail need to be plucked out one or two at a time.

Because feathers and down float in every direction when waterfowl are dry picked, choose a setting that is free from drafts. Picking into a large plastic garbage bag and covering the floor around the plucking area with newspapers are helpful in keeping feathers under control.

Scald Picking

The most common method for defeathering ducks is to scald them prior to picking. The only equipment needed is a large container in which to submerge an entire duck, and a forked branch (or a 1″ × 2″ × 24″ board with a large nail driven in at an angle near one end) which is used to hold the bird under water.

To scald, hold the duck by its feet and wedge the neck into the fork of the stick. Then dip the bird up and down in water that is 125° to 145° F., making certain that the water penetrates through to the skin. To improve the wetting ability of the scalding water, a small amount of detergent can be added.

The scalding time for ducks varies from one to three minutes, depending on the age of the bird and the temperature of the water. Mature ducks require longer and hotter scalds than ducklings. If you find that the feathers are still difficult to remove after the initial scald, the bird can be redipped. Overscalding causes the skin to tear easily and discolors the carcass with dark blotches.

Ducks should be picked immediately after scalding, starting with the wing and tail feathers. Because the feathers are going to be hot, have a bucket of cold water nearby to dip your hands into occasionally.

If your first attempts at scald picking poultry do not produce carcasses that are as attractive as those that are processed commercially, don't get discouraged. Your results should improve once you've gained a little experience.

Feathers from scalded ducks are of good quality when handled correctly (see Care of Feathers, page 120).

Wax Picking

A popular variation of the scalding method is to dip rough-picked ducks in hot wax. Even when birds with a moderate number of pin feathers are butchered, this procedure can produce clean carcasses in a short time. Paraffin or a mixture of one part beeswax to one part paraffin can be used. There are also products such as Dux-Wax which are made specifically for this purpose and are available from poultry supply distributors. (See Appendix H, page 164.)

When wax is used, ducks are scalded without detergent in the water, and rough-picked by removing the large tail and wing feathers and 50 to 90 percent of the body plumage. Prior to scalding, place a container of solidified wax in a large receptacle of hot water where it is melted and heated to a temperature of 150° to 165° F. *Extreme care* must be taken when working with hot wax to avoid burns and fires.

Rough-picked ducks are partially dried and then dipped into the hot wax several times. Spray with cold water or wait long enough between each dunking to allow sufficient congealing to build up a good layer of wax. If only a few birds are being dressed, you may find it simpler to melt a small container of wax to be poured over the carcasses.

Submerging the waxed duck in cold water causes the wax to harden and grip the feathers. The wax and feathers are then stripped off together, resulting in a finished product that is clean and attractive. Birds that are extra pinny can be rewaxed.

Used wax can be recycled by melting it and skimming off the feathers and scum, and boiling out any existing water.

SINGEING

After ducks are picked, long hair-like filament feathers usually remain. The simplest way to remove these filoplumes is to pass the carcasses quickly over a flame, being careful not to burn the skin. A jar lid with a thin layer of rubbing alcohol in the bottom gives the best flame I know of for singeing. Alcohol burns tall, cleanly and odor-free. Newspapers (do not use colored sheets) loosely rolled into a hand torch, gas burners and candles can also be used.

SKINNING

Ducks can be skinned rather than picked. Some advantages of this technique are that ducks with pin feathers can be dressed as easily as those in full-feather and some people find skinning less time-consuming than picking. Because skin is composed largely of fat, skinning significantly reduces the fat content of the dressed duck.

The major drawbacks of this method are that skinned carcasses lose much of their eye appeal when roasted whole, special precautions must be taken in cooking the meat to prevent dryness, some flavor is lost and a higher percentage of the bird is wasted.

To prepare a duck for skinning, remove its head, feet and the last joint of each wing. With the bird resting on its back, slip the blade of a small, sharp knife under the skin of the neck, and slit the skin the length of the body, cutting around both sides of the vent. The final step is to peel the skin off, which requires a good deal of pulling. Over stubborn areas, a knife is needed to trim the skin loose.

EVISCERATING

Ducks can be drawn immediately after they have been defeathered, or they can be chilled in ice water for several hours or hung in a cool (33° to 36° F.) location to ripen for six to twenty-four hours. Chilling the carcasses first has the advantage of making the cleaning procedure less messy, while aging before eviscerating produces stronger-flavored meat, which is preferred by some people.

With the bird resting on a clean, smooth surface (we use our dishrack drainboard), remove the feet, neck and oil gland, making certain that *both* yellowish lobes of the oil gland are cut out. Then make a shallow three-inch-long horizontal incision between the end of the breastbone and the vent, being careful not to puncture the intestines that lie just under the skin.

Through the incision insert your hand into the body cavity and gently loosen the organs from the inside walls of the body, and pull them out. Cut around the vent to disconnect the intestines from the carcass. The gizzard, heart and liver should be cut free and set aside before the unwanted innards are discarded.

Cutting out the oil gland

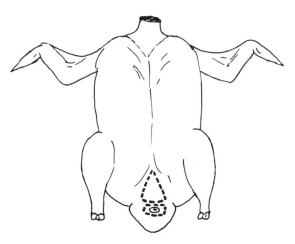

Cutting bird for eviscerating
. . . above and around vent

The esophagus (ducks do not have true crops) and windpipe are well-anchored in the neck and require a vigorous pull to remove them. The pink, spongy lungs are located against the back among the ribs and can be scraped out with the fingers if you desire.

To clean gizzards, cut around their outside edge and then pull the two halves apart. The inner bag with its contents of feed and gravel can then be peeled away and discarded. The final step is to rinse the muscular organ with water.

The gall bladder, a small green sac, is tightly anchored to the liver and should be removed *intact* since it is filled with bitter bile. A portion of the liver must be cut off along with the gall bladder so bile does not spill onto edible meat.

Unwanted feathers and body parts make excellent fertilizer and should be buried near a tree or in the garden. *Do not feed raw innards to cats and dogs.* If you give uncooked entrails to your pets, they can develop a taste for poultry and may kill birds to satisfy their cravings.

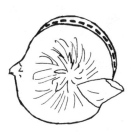

Cutting open gizzard

Taking out contents

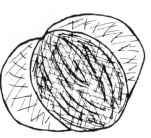

Cleaned gizzard

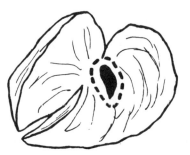

*Incision to remove
green gall bladder from liver*

COOLING THE MEAT

After all the organs have been removed, thoroughly wash the carcass and chill it to a temperature of 34° to 40° F. as soon as possible. If meat is cooled slowly, bacteria may grow, causing spoilage and unpleasant flavor. Poultry can be chilled in ice water or air-cooled by hanging the carcasses in a refrigerator or in a room with a temperature of 30° to 40° F.

PACKAGING AND STORING MEAT

After the body heat has dissipated from the carcasses, they should be sealed in airtight containers. If the meat was chilled in water, it should be allowed to drain for ten to twenty minutes before it is packaged. To retain the highest quality in meat that is going to be frozen, suck the air from plastic bags with a straw or vacuum cleaner before sealing. To produce the tenderest meat, poultry must be aged twelve to thirty-six hours at 33° to 40° F. before it is eaten or frozen.

CARE OF FEATHERS

Duck feathers are a valuable by-product of butchering. If you plan to save the feathers, keep the down and small body feathers separate from the large stiff plumes of the wings, tail and body as the slaughtered birds are being picked.

When ducks are scalded prior to picking, the feathers need to be washed with a gentle detergent, rinsed thoroughly in warm water, and spread out several inches thick on a clean, dry surface or loosely placed in cloth sacks and hung in a warm room. Stir the wet feathers daily to fluff them and insure rapid drying. Once they are well-dried, feathers can be bagged and stored in a clean, dry location. (See Appendix F, page 157, for instructions on how to use feathers.)

BUTCHERING CHECK LIST

1. Remove the ducks' feed four to six hours before they are killed.
2. Sharpen all cutting tools.
3. When catching ducks prior to butchering, keep them calm and handle them carefully to avoid discolored meat due to bruises.
4. Hang slaughtered ducks in killing cones or by their feet to avoid bruising or soiling the meat and to insure thorough bleeding.
5. Remove feathers as soon as possible after birds are bled.
6. Singe off filament feathers.
7. Trim out the oil gland from the base of the tail.
8. Cut off feet and shanks at hock joints; remove neck.
9. Make a horizontal incision between the vent and the end of the breastbone, and gently lift out innards.
10. Set aside the heart, gizzard and liver before disposing of unwanted innards.
11. Pull out the windpipe and esophagus from the neck/chest area.
12. Extract the lungs from between the ribs.
13. Rinse the carcass thoroughly with clean, cold water.
14. Clean the gizzard and carefully remove the gall bladder from the liver.
15. Chill dressed duck to an internal temperature of 40° F.
16. Bury unwanted body parts, entrails and feathers in the garden or near a tree.
17. Age the meat for twelve to thirty-six hours at 33° to 40° F. prior to cooking or freezing.
18. Package and freeze the meat, or enjoy a festive banquet of roast duck.

CHAPTER 11

Health and Physical Problems

Ducks in small flocks are rarely bothered by poor health. When problems do arise, they can often be traced to diet deficiencies, although stagnant drinking water, medicated feeds, moldy bedding, or overcrowded and filthy conditions can also be the source of illness.

If a problem does appear and you are unable to diagnose the trouble, advice should be sought *at once* from an experienced duck breeder, veterinarian or animal diagnostic laboratory. Waiting to see if an ailment will cure itself can be fatal for the duck.

Sick or injured birds should be isolated in a dry, clean pen, and provided a balanced diet and clean drinking water. Remove dead birds *immediately* to avoid the possibility of attracting predators and, in the case of communicable disease, to curb the likelihood of infecting healthy ducks. Bury carcasses deep enough that they will not be dug up by scavengers.

The following ailments are the ones that are the most common in small duck flocks. Nearly all of these conditions can be diagnosed and treated at home.

BOTULISM

Cause. This deadly food poisoning is caused by a toxin commonly found in decaying animal and plant matter. Botulism most frequently strikes when the weather has been dry and water levels in ponds and lakes drop, leaving decaying plants and fish exposed where ducks find and eat them. Ducks can contract botulism if they are given spoiled canned foods from the pantry.

Symptoms. A few hours after eating poisoned food, birds may lose control of their leg, wing and neck muscles. Body feathers frequently loosen and are easily extracted. Ducks that are swimming when paralysis of the neck develops often drown before they are able to climb out on land. Dying birds may slip into a coma several hours before expiring. Botulism kills in three to twenty-four hours, although in mild cases, birds may recover in several days.

Treatment. Confine ill ducks to a clean yard and provide fresh drinking water *immediately.* Assist any birds that cannot drink on their own. Giving a laxative consisting of a half-cup molasses in each gallon of drinking water for approximately four hours, or individual doses of one-half to one teaspoon of castor oil, often helps. Epsom salts—one pound per five gallons of drinking water—has also been recommended as an effective treatment. A vaccine has been developed, but is rather expensive and often difficult to obtain on short notice. Every effort must be made to locate the source of the poisoning so that future problems can be avoided.

Prevention. Bury carcasses of dead animals and clean up rotting vegetation. Do not let your ducks feed in stagnant bodies of water during dry periods.

EVERSION OF THE OVIDUCT

Cause. This ailment can be the result of a hen straining to lay an unusually large egg or can be caused by oviduct muscles being weakened due to premature or high egg production or obesity. Unfortunately this disorder occurs most frequently among the best layers.

Symptoms. A hen with this problem is easily identified by her droopy appearance and the expelled oviduct protruding from her vent.

Treatment. An ailing hen can be saved only if she is discovered within several hours after the oviduct is dislodged and *prompt* action is taken. Even then, her chances of healing are marginal.

To treat, wash the oviduct thoroughly with clean, warm water. Apply an antiseptic ointment, then gently push the organ back into place. Confine the hen in a clean pen for several days. To give the muscles a chance to heal, the hen should be fed small quantities of feed that will discourage laying.

Hens that are not valuable enough for treatment should be killed immediately and dressed for meat or buried. Left unattended, they will suffer a slow, painful death.

Prevention. Do not push pullets to begin laying before they are mature (twenty to twenty-four weeks) and make sure hens are not excessively fat. To keep the oviduct of high-producing hens healthy and lubricated, mix cod liver oil (one teaspoon per bird) with the feed one day a week.

FEATHER EATING

Cause. Cannibalism is most prevalent among ducklings, especially Muscovies, that are brooded artificially. It can be the result of high brooding temperature, excessive light, overcrowding, an unbalanced diet, insufficient quantities of feed or lack of roughage.

Symptoms. Birds pull out and eat one another's feathers.

Treatment. At the first sign of feather eating, check brooder temperature, reduce the intensity of light (using blue bulbs often helps), and provide ducklings with sufficient space, a balanced diet, adequate quantities of feed and tender green foods such as grass, clover and dandelions.

Prevention. Correct brooding temperature; dim, colored lights; sufficient floor space; a balanced diet and the daily feeding of greens eliminate this disgusting problem.

FOOT PROBLEMS

Cause. The bottoms of the feet of waterfowl are more tender than those of land-dwelling birds such as chickens. Foot problems can be the result of ducks bruising the pads of their feet or from spending much of their time on dry, hard surfaces without having access to bathing water. However, the most common cause, although not usually recognized, is a dietary deficiency in biotin, pantothenic acid or one of the other vitamins or minerals that are important in maintaining healthy skin.

Symptoms. Large corns and rough calluses develop on bottoms of feet. In severe cases, deep, bleeding cracks are evident and/or the pads of the feet swell and become infected (commonly known as bumble foot), causing the bird to go lame.

Treatment. If it is suspected that the problem is the result of a deficient diet, supplement the bird's rations with a vitamin premix or feedstuffs, (such as brewer's dried yeast, whey, dried skim milk or alfalfa meal), that are rich in vitamin A, biotin and pantothenic acid.

If the foot is inflamed, open the callus with a sharp, sterilized instrument, remove the pus or core, and disinfect the incision with rubbing alcohol. Place the duck in a clean pen that is bedded down with a thick layer of straw. Provide a balanced ration, green feed and at least a small container (a dish pan will suffice) of bathing water. The daily washing, disinfecting and application of a moisturizer may facilitate healing.

Rough, calloused feet such as these are often the result of a diet deficient in biotin, pantothenic acid and/or riboflavin.

Prevention. Ducks who receive a balanced diet seldom develop foot problems. Access to succulent pasture and bathing water reduces its occurrence.

HARDWARE DISEASE

Cause. Nails, bits of wire or pieces of string are swallowed by birds. When ingested, these objects often impact or puncture some portion of the digestive system.

Symptoms. Birds slowly lose weight, stop eating and sit with eyes partially closed, apparently in severe pain. When a post-mortem is performed, the hardware is often found lodged in the esophagus or gizzard.

Treatment. There is no practical remedy for hardware disease, unless the problem is a crop (esophagus) impaction, often called pendulous crop, caused by ingestion of string or tough green feed. In this case the crop can be slit open and the object removed. Afterwards, limit food intake until muscle tone returns.

Prevention. Never leave nails, wire or string where birds can reach them. Unfortunately, it takes only *one* misplaced nail to cause the death of a valuable bird. Whenever hardware is being used in an area to which birds have access, every effort should be made to retrieve bits and pieces that are dropped.

MALNUTRITION

Cause. Malnutrition can be caused either by an incomplete diet or an insufficient quantity of feed. It occurs most frequently when birds are raised in buildings or grassless yards and are fed nothing but grains or inadequate chicken rations.

Symptoms. In ducklings, malnutrition can be detected by stunted growth, wide variation in body size, retarded feather development, leg weakness and emaciated birds with little resistance to disease or parasites. Adult ducks that are undernourished produce poorly, have rough-looking feathers, may be thin and are susceptible to disease and parasites.

Treatment. Provide an ample quantity of food that supplies a balanced diet. (See Nutrition, Chapters 8 and 9.)

Prevention. Do not try to save money by starving ducks or by using a cheap feed that does not furnish a complete diet. If you are feeding grains, make sure the birds have access to a plentiful supply of grass and insects, or a protein, vitamin and mineral concentrate.

MEDICATION POISONING

Cause. Ducks are frequently poisoned when fed medicated chicken or turkey feeds. The problem seems to stem from the fact that in proportion to their body weight, ducks eat more feed than land fowl, getting an overdose of the medication. If ingested in the correct dosages, these drugs do not affect ducks any differently than chickens or turkeys.

Symptoms. Birds (usually ducklings that are brooded in close confinement) lose their appetites, become weak, have stunted growth or die suddenly.

Treatment. At the first sign of any of these symptoms, switch to a nonmedicated ration. If medicated feeds must be used, dilute the potency of the drug by providing an abundant supply of succulent greens and mix rolled, cracked or small, whole grains with the feed.

Prevention. Unless you have no other choice, do not use medicated rations, particularly if the birds are not able to forage some of their own nourishment.

NIACIN DEFICIENCY

Cause. Ducklings are maintained on a diet deficient in niacin.

Symptoms. Birds develop weak or bowed legs, and often show stunted growth. (Rickets, sometimes confused with niacin deficiency symptoms, is caused by a vitamin D₃ deficiency.)

Treatment. Ducks exhibiting mild symptoms of a niacin deficiency can usually be cured by the immediate addition of a niacin supplement (see Niacin Requirements, page 78) to their feed or drinking water. In severe cases, ducklings may become so crippled that they are worthless and must be destroyed.

Prevention. When allowed to forage in pastures and bodies of water where there is an abundant supply of insects, ducklings are seldom bothered by this malady. If raised in close confinement, they must be fed a niacin-rich diet.

PARASITES, EXTERNAL

Cause. Lice and mites seldom multiply on waterfowl, particularly those with access to bathing water, in large enough numbers to cause ill effects. However, ducklings that are hatched and brooded by chicken or turkey hens can harbor heavy infestations of external parasites.

Symptoms. Parasites are visible on birds if you look closely, particularly on the head and wings, and around the vent and oil gland. Infected ducks lose weight, grow slowly and are anemic.

Treatment. Use olive oil, pulverized dried tobacco leaves or a commercial preparation such as Sevin Dust or Malathion. To be effective, these products need to be worked into the feathers of the head, neck, wings, upper tail and vent. When you are dusting with an insecticide, be *extremely* careful not to contaminate water or feed with the poison.

Prevention. Provide bathing water or treat birds before lice or mites are present in large enough numbers to be harmful. Turkey and chicken hens used as foster mothers for ducklings should *always* be treated for lice.

PARASITES, INTERNAL

Cause. Healthy ducks consume such large volumes of water that worms are usually flushed out of their digestive systems faster than the parasites can reproduce. Worms are usually a problem only when ducks have access to stagnant water and crowded ponds or small streams, or when they are forced to live in a filthy environment.

Symptoms. An infestation of worms retards growth, lowers feed conversion and reduces egg production. In severe cases, these symptoms are accompanied by a loss of body weight and eventually death.

Treatment. Poultry worm medications that are added to the drinking water are readily available. Some people have suggested diatomaceous earth and chewing tobacco as alternative remedies. Because many of the wigglers will be passed alive, the ducks should be closed in a small pen for twenty-four hours after being wormed. The bedding should then be cleaned up and disposed of, preferably by burning.

Prevention. Provide fresh drinking water and reasonably sanitary living quarters. As a precaution, birds can be routinely wormed once or twice yearly.

PHALLUS PROSTRATION

Cause. A wild drake normally pairs off with a single hen and is sexually active for a relatively short period of time each spring. Under domestication, males are frequently mated with two to seven hens and breed over an extended season. This unnaturally long mating season, or possibly a genetic weakness, occasionally causes a drake to lose the ability to retract his penis.

Symptoms. The penis—a 1½-inch long, coiled organ—protrudes from the bird's vent.

Treatment. Under most circumstances, a drake with this disability should be killed *immediately*. If the bird is valuable and his problem is discovered before the penis has become infected or dried out, there is a possibility of recovery. To treat, wash and then disinfect the organ with an antibacterial ointment and isolate the drake where he has clean swimming water. Keep the bird in isolation and apply the ointment daily until he is fully recovered, which may require several weeks or longer.

Prevention. The best safeguard is to have an extra drake so that if one develops this prostration, you'll have a backup male for breeding purposes.

SLIPPED WING

Cause. A genetic weakness, diet deficiency or extremely fast growth can cause slipped wings.

Symptoms. The tip of one or both wings folds on the outside of the pinion, rather than resting smoothly against the bird's sides.

Treatment. When prompt action is taken, it is sometimes possible to repair slipped wings of ducklings by folding the feathered limb in the correct position and taping it shut. Because the wing will become stiff from disuse, the bandage should not be left on for more than four or five days at a time.

Prevention. The occurrence of slipped wing can be reduced by using only normal-winged breeding stock, and by making sure ducklings consume a

A Black Swedish duck with slipped wings.

balanced diet. Birds with this deformity are fine for meat or the production of eating eggs.

SPRADDLED LEGS

Cause. Spraddled legs are usually the result of smooth incubator trays or brooder floors where ducklings have poor footing, but this problem can also be the result of a birth defect.

Symptoms. Birds are crippled by weak legs that slide out from under them as they try to walk.

Treatment. Severe cases of spraddled legs are difficult to correct. Ducklings with mild cases are sometimes rehabilitated by placing them on a rough surface that gives good footing—wood excelsior pads as used in chick boxes, grass turf or coarse burlap are excellent for this purpose. Tying a short piece of yarn between the duck's legs for two or three days is also helpful. When hobbles are used, don't tie them so tightly that blood circulation to the feet is restricted.

Prevention. Slick surfaces on which ducklings walk must be covered with a

129

Spraddled legs

Hobbled legs

rough material such as wire hardware cloth or burlap for the first week after the birds hatch.

STICKY EYE

Cause. In their natural habitat, ducks consume foods high in vitamin A, pantothenic acid and biotin, and have access to clean bathing water. Under domestication, waterfowl are often raised on diets deficient in vitamins and are supplied water in shallow containers which do not permit the birds to rinse their eyes. The result is that ducks raised in confinement are susceptible to ophthalmia, a low-grade infection in one or both eyes which is referred to as sticky eye.

Symptoms. A yellowish discharge mats down the feathers around the eye and may cause the eyelids to stick shut. Sticky eye is most prevalent among ducklings that are raised indoors.

Treatment. As soon as opthalmia is detected, ducks should be fed fresh greens and given a vitamin supplement until the problem clears up. Providing clean drinking water deep enough for the birds to submerge their heads, and the daily application of a medicated eye ointment speed the recovery in some cases. If sticky eye is not cured promptly, the infection can linger for the duration of the bird's life and cause blindness.

Prevention. If ducklings are raised in confinement and not fed a ration formulated specifically for waterfowl, fortify their diets with a vitamin premix or

Adult drake with a chronic case of sticky eye

other substances (such as alfalfa meal, brewer's dried yeast or green feed) that are high in vitamin A, biotin and pantothenic acid. Disinfect water containers weekly.

APPENDICES

Formulating Duck Rations

Feed mills often have someone on their staffs who is trained in formulating livestock and poultry feeds. However, with the aid of a hand calculator and by using the information in this appendix, you should be able to devise good duck rations that utilize readily available local ingredients. While ration formulation requires several hours of work, it can save you considerable money on your feed bill.

The purpose of this section is to present the basic nutrient requirements of ducks and explain how to formulate rations. If you are interested in the whys of nutrition, see Appendix I, page 166, for sources of information.

USING THE FORMULA CHART

The formula chart (see Table 22) is a convenient and orderly means by which rations can be computed. Ingredients and their values for protein, energy, fiber, etc., are taken from Table 21 (or similar chart) and written or typed into the proper columns of the formula chart. The percentage (which should be expressed as a decimal, e.g., 73 percent wheat ÷ 100 = .73) at which each ingredient is going to be included in a ration is then multiplied times the various nutrients of the ingredient.

Once the protein values have been calculated for each ingredient, they should be added up to see if the protein content of the ration compares favorably with the recommendations given in Table 20. Once the proper protein content has been obtained, the values for energy, fat, fiber, minerals, vitamins and amino acids should be computed and compared with Table 20.

CONVERTING KILOGRAMS TO POUNDS

In many publications containing information showing the average composition of feedstuff for poultry (similar to Table 21), the values for the various nutrients are given in kilograms. If you'd rather work in the American Standard rather than the Metric system of weights and measures, a conversion must be made. One kilogram equals 2.2 pounds, so to convert amounts per kilogram to amounts per pound, *divide* by 2.2. Example: Wheat has approximately 3,120 kcal/kg of metabolizable energy, which if divided by 2.2 gives 1418 kcal/lb of ME. (If you want to work in the metric system, the values in Tables 20 and 21 must be converted from pounds to kilograms by multiplying by 2.2.) Those values that are given in percentages, (such as protein, calcium, phosphorus, etc.), *remain the same* no matter which system is used.

COMPUTING THE CALORIE:PROTEIN RATIO

Birds eat primarily to meet their energy needs. To guard against over-or-under-consumption of protein by ducks, rations must have the proper calorie-to-protein ratio (see Table 20). To compute this ratio, divide the metabolizable energy (kcal per pound of feed) by the protein percentage of the ration (e.g., 1278 kcal of energy ÷ 15.8 percent protein = 80.9).

THE PHOSPHORUS:CALCIUM RATIO

In order for ducks (particularly young ones) to utilize phosphorus and calcium effectively, these two minerals must be included in feeds in the proper proportions. An imbalance of phosphorus and calcium can result in stunted growth, rickets and other bone deformities, and in acute cases, lead to death.

For ducklings and non-producing adults, the correct ratio of total phosphorus to calcium is in the range of 1:1 to 1:1.5 (P:Ca), with 1:1.2 considered ideal. For laying birds, the ratio should fall between 1:4 and 1:5.

To compute the phosphorus:calcium ratio of a ration, divide the amount of calcium by the quantity of total phosphorus. Example: A starting ration contains .80 percent calcium and .65 percent phosphorus (.80 ÷ .65 = 1.2 which is equal to a P:Ca ratio of 1:1.2).

USING A PROTEIN CONCENTRATE

In many instances, the simplest method for preparing a balanced and economical ration at home is by mixing a commercial protein concentrate with locally grown grains. If fortified with additional niacin, concentrates formulated for chickens normally work satisfactorily for ducks as well.

To calculate the proportions of concentrate to grains that should be used, Pearson's Square Method can be employed. Example: You have a concentrate that is 40 percent protein and want to blend it with soft wheat (approximately 10 percent protein) to get a 16 percent protein laying ration.

10% protein in wheat	24 parts of wheat
16% protein desired	
40% protein in the concentrate	6 parts of concentrate

To obtain the correct proportions of wheat and concentrate needed, subtract the small number from the larger on each diagonal (i.e., 16 − 10 = 6 parts concentrate and 40 − 16 = 24 parts wheat). Then divide the parts of each by the total number of parts and multiply by 100 to get the percentage at which each ingredient should be used:

6 + 24 = 30 total parts
6 ÷ 30 = .2 × 100 = 20 percent of protein concentrate needed
24 ÷ 30 = .8 × 100 = 80 percent of wheat needed

RECOMMENDED NUTRIENT LEVELS
FOR COMPLETE DUCK RATIONS (PELLETED)[1]

(Expressed as a percentage or amount per pound of feed)

Nutrient	Starter 0-2 Wks.	Grower-Finisher 2-7 Wks.	Breeder Devel-oper	Breeder Layer	Holding Ration
*Metabolizable energy (kcal/lb)[2]	1400	1400	1400	1300	1400
*Energy:Protein ratio	70	88[3]	88	81	100
*Fat, minimum (%)	3.0	2.5	2.0	2.0	1.5
*Fiber, maximum (%)	4	5	6	6	7
*Protein, minimum (%)	20	16	16	16	14
Amino Acids:					
Arginine (%)	1.25	1.00	1.00	0.95	0.83
Glycine + serine (%)	1.12	0.90	0.90	0.85	0.74
Isoleucine (%)	0.44	0.35	0.35	0.30	0.26
Histidine (%)	0.88	0.70	0.70	0.65	0.57
Leucine (%)	1.63	1.30	1.30	1.20	1.05
*Lysine (%)	0.94	0.75	0.75	0.70	0.62
*Methionine + cystine (%)	0.75	0.60	0.60	0.55	0.48
*Methionine (%)	0.44	0.35	0.35	0.30	0.26
Phenylalanine + tyrosine (%)	1.63	1.30	1.30	1.20	1.05
Phenylalanine (%)	0.88	0.70	0.70	0.65	0.57
Threonine (%)	0.75	0.60	0.60	0.55	0.48
Tryptophan (%)	0.24	0.19	0.19	0.17	0.15
Valine (%)	1.00	0.80	0.80	0.75	0.66
Vitamins (fat soluble):					
*Vitamin A (IU)	4750	3250	3250	4750	3250
*Vitamin D₃ (ICU)	500	400	400	400	400
*Vitamin E (IU)	12	8	8	15	8
*Vitamin K (mg)	1.0	0.5	0.5	1.0	0.5

[1]Nutrient levels shown in this table apply only to the energy level specified. For rations containing different energy concentrations, compensations in the quantities of amino acids, vitamins and minerals should be made to guard against over- or under-consumption of these nutrients. If your ration has a lower energy level than given in this table, the amounts of the other nutrients can likewise be reduced to avoid waste. Example: Your ration has an energy level of 1260 kcal/lb, which is 90% of 1400 kcal/lb (1260 ÷ 1400 = .9 × 100 = 90%). Therefore, all nutrients can be reduced by approximately 10% and the birds will still ingest the proper quantities of all nutrients.

[2]The energy concentration given is only an example. The energy concentration of rations may vary from 1000 to 1500 kcal/lb, provided the concentration of each nutrient per unit of energy remains the same. The energy level chosen should be the one at which least cost per unit of energy is achieved.

[3]To produce ducklings with less fat, the energy:protein ratio can be lowered to 65 or less.

[4]Most of the niacin in plants is unavailable to birds.

[5]Rations based on wheat, milo or barley are often deficient in linoleic acid, necessitating the addition of an ingredient rich in this fatty acid, such as one of the vegetable oils.

Nutrient	Starter 0-2 Wks.	Grower-Finisher 2-7 Wks.	Breeder Developer	Breeder Layer	Holding Ration
Vitamins (water soluble):					
Thiamine (mg)	1.25	1.25	1.25	1.25	1.25
*Riboflavin (mg)	3.75	2.25	2.25	3.75	2.25
*Pantothenic acid (mg)	7	6	5	7	5
*Niacin, available (mg)[4]	35	30	25	35	25
Pyridoxine (mg)	1.45	1.45	1.45	1.65	1.45
Biotin (mg)	0.09	0.06	0.06	0.09	0.06
*Choline (mg)	800	700	450	450	450
Folacin (mg)	0.35	0.20	0.20	0.25	0.20
*Vitamin B_{12} (mg)	0.006	0.005	0.004	0.004	0.004
Linoleic acid (%)[5]	01.0	0.8	0.8	1.0	0.8
Minerals:					
*Calcium, minimum (%)	0.70	0.60	0.60	2.75	0.60
*Calcium, maximum (%)	1.00	1.00	1.00	3.00	1.00
Phosphorus, total (%)	0.65	0.60	0.60	0.65	0.60
*Phosphorus, available (%)	0.40	0.35	0.35	0.40	0.35
Potassium (%)	0.4	0.4	0.3	0.2	0.2
*Sodium (%)[6]	0.17	0.14	0.14	0.14	0.14
Chlorine (mg)[6]	400	400	400	400	400
Copper (mg)	2.0	1.5	1.5	2.0	1.5
Iodine (mg)	0.20	0.20	0.20	0.18	0.18
Iron (mg)	40	20	20	30	20
Magnesium (mg)	275	275	275	275	275
*Manganese (mg)	25	25	25	20	20
Selenium (mg)[7]	0.06	0.06	0.06	0.06	0.06
*Zinc (mg)	15	15	15	15	15

[6]These recommendations are for birds receiving low salt drinking water. The salt content of rations should be reduced if drinking water contains above average levels of sodium chloride.

[7]Grains grown in areas where the soil is of volcanic origin, often are deficient in selenium.

*Denotes the nutrients that are of greatest concern when formulating rations. Usually, if these are balanced, the others will be satisfactory.

Note: Information in this table was adapted from Extension Stencil #25 (1969) "Duck Rations," and updated information, (1972 and 1973) with permission of Cornell University Duck Research Laboratory; and from Nutrient Requirements of Poultry, 7th revised ed., (1977), pages 30 and 33, with the permission of the National Academy of Sciences, Washington, D.C.

KEY: IU = International Units
ICU = International Chick Units
mg = milligram
kcal = kilocalories

TABLE 21

AVERAGE COMPOSITION OF SOME COMMONLY USED FEEDS FOR POULTRY

Line No.	Feedstuff	Pro-tein %	Energy kcal/lb	Crude Fat %	Crude Fiber %	Cal-cium %	Total Phos-phorus %	Avail. Phos-phorus %	So-dium %	Man-ganese mg/lb
	Alfalfa meal, dehydrated									
1	17% Protein	17.5	622.73	2.0	24.1	1.44	0.22	0.22	0.12	13.6
2	20% Protein	20.0	740.91	3.6	20.2	1.67	0.28	0.28	0.13	19.2
3	Barley	11.6	1200.00	1.8	5.1	0.03	0.36	0.11	0.04	—
4	Barley, Pacific Coast	9.0	1190.91	2.0	6.4	0.05	0.32	0.10	0.02	7.4
5	Corn, yellow	8.8	1559.09	3.8	2.2	0.08	0.28	0.08	0.02	2.3
6	Corn, Gluten meal, 41%	41.0	1336.36	2.5	7.0	0.23	0.55	0.16	0.07	4.0
7	Cottonseed meal, solvent	41.4	1090.91	0.5	13.6	0.15	0.97	0.35	0.04	9.1
8	Dicalcium phosphate	—	—	—	—	21.00	18.50	18.50	1.20	54.5
9	Distiller's dried solubles (corn)	28.5	1331.82	9.0	4.0	0.35	1.33	0.39	0.26	33.5
10	DL-Methionine, 98%	—	—	—	—	—	—	—	—	—
11	Feather meal, hydrolyzed	86.4	1072.73	3.3	1.0	0.33	0.55	0.55	0.71	9.5
12	Fish meal, herring	72.3	1450.00	10.0	0.7	2.29	1.70	1.70	0.61	2.1
13	Fish meal, menhaden	60.5	1281.82	9.4	0.7	5.11	2.88	2.88	0.41	15.0
14	Limestone, ground	—	—	—	—	38.00	—	—	0.06	112.5
15	Meat and bone meal	50.4	890.91	8.6	2.8	10.10	4.96	4.96	0.72	6.4
16	Meat meal	54.4	909.09	7.1	8.7	8.27	4.10	4.10	1.15	4.4
17	Oats	11.4	1159.09	4.2	10.8	0.06	0.27	0.08	0.08	19.6
18	Oats, West Coast	9.0	1186.36	—	11.0	0.08	0.30	0.09	—	17.3
19	Oyster shell	—	—	—	—	37.26	0.07	0.07	0.20	58.0
20	Salt, iodized	—	18.18	—	—	—	—	—	38.91	—
21	Sorghum, grain (milo)	8.9	1531.82	2.8	2.3	0.03	0.28	0.09	0.04	6.2
22	Soybean meal, solvent 50%	50.8	1131.36	1.1	2.9	0.29	0.65	0.20	0.01	20.9
23	Soybean meal, solvent 44%	44.0	1013.64	0.8	7.3	0.29	0.65	0.20	0.26	13.3
24	Soybean meal, dehulled	48.5	1109.09	1.0	3.9	0.27	0.62	0.19	0.25	19.5
25	Wheat, bran	15.7	590.91	3.0	11.0	0.14	1.15	0.35	0.05	51.5
26	Wheat, hard	14.1	1272.73	1.9	2.4	0.05	0.37	0.11	0.04	14.5
27	Wheat, middlings	16.0	818.18	3.0	7.5	0.12	0.90	0.28	0.12	53.6
28	Wheat, soft	10.2	1418.18	1.8	2.4	0.05	0.31	0.10	0.04	10.8
29	Whey, dried	12.0	863.64	0.8	0.2	0.97	0.76	0.76	0.48	2.8
30	Yeast, brewer's, dried	44.4	904.55	1.0	2.7	0.12	1.40	0.41	0.07	2.4

TABLE 21-Continued

Line No.	Zinc mg/lb	Vit. A IU/lb	Vit. D_3 ICU/lb	Vit. E mg/lb	Vit. K mg/lb	Ribo-flavin mg/lb	Panto-thenic Acid mg/lb	Total Niacin mg/lb	Choline mg/lb	Vit. B_{12} mg/lb	Ly-sine %	Methi-onine %	Cys-tine %
1	10.9	83909	—	56.82	4.0	6.2	11.4	17.3	637	.0018	0.73	0.23	0.20
2	11.4	101000	—	65.46	4.7	6.9	15.5	18.2	645	.0018	0.87	0.31	0.25
3	7.7	—	—	9.09	—	0.8	3.6	25.0	450	—	0.40	0.17	0.19
4	6.8	1318	—	9.09	—	0.7	3.2	21.8	470	—	0.29	0.13	0.18
5	4.5	11955	—	10.00	—	0.5	1.8	10.9	282	—	0.24	0.20	0.15
6	9.1	—	—	9.09	—	0.8	4.5	22.7	421	—	0.78	1.03	0.65
7	37.3	—	—	—	—	1.8	3.2	18.2	1333	—	1.71	0.52	0.64
8	13.6	—	—	—	—	—	—	—	—	—	—	—	—
9	38.6	500	—	18.18	—	7.7	9.5	52.7	2201	—	0.90	0.50	0.40
10	—	—	—	—	—	—	—	—	—	—	—	98.00	—
11	24.5	—	—	—	—	1.0	4.5	12.3	405	.0355	1.67	0.42	4.00
12	60.0	—	—	10.00	—	4.5	7.7	42.3	2412	.1832	5.70	2.10	0.72
13	66.8	—	—	3.18	—	2.2	4.1	25.0	1389	.0473	4.83	1.78	0.56
14	—	—	—	—	—	—	—	—	—	—	—	—	—
15	42.3	—	—	.46	—	2.0	1.9	20.9	907	.0318	2.60	0.65	0.25
16	46.8	—	—	.46	—	2.5	2.3	25.9	944	.0309	3.00	0.75	0.66
17	7.7	—	—	9.09	—	0.5	—	5.5	430	—	0.50	0.18	0.22
18	—	—	—	9.09	—	0.5	5.9	6.4	436	—	0.40	0.13	0.17
19	—	—	—	—	—	—	—	—	—	—	—	—	—
20	—	—	—	—	—	—	—	—	—	—	—	—	—
21	6.4	—	—	5.46	—	0.5	5.5	18.6	205	—	0.22	0.12	0.15
22	20.5	—	—	2.23	—	1.4	6.2	9.9	1265	—	3.19	0.74	0.83
23	12.3	182	—	.96	—	1.3	7.3	13.2	1270	—	2.93	0.65	0.69
24	20.5	—	—	1.50	—	1.3	6.8	10.0	1241	—	3.18	0.72	0.73
25	60.5	—	—	6.14	—	2.1	14.1	84.5	855	—	0.59	0.17	0.25
26	14.1	—	—	5.73	—	0.6	4.5	21.8	495	—	0.40	0.19	0.26
27	68.2	—	—	18.41	—	1.0	5.9	44.5	654	—	0.69	0.21	0.32
28	12.7	—	—	6.00	—	0.5	5.0	25.9	455	—	0.31	0.15	0.22
29	1.4	—	—	.09	—	12.3	20.0	4.5	622	.0105	0.97	0.19	0.30
30	17.7	—	—	—	—	16.8	49.5	203.6	1811	—	3.23	0.70	0.50

Note: Information in this table was reproduced from *Nutrient Requirements of Poultry*, 7th revised ed., (1977), pages 42–45, with the permission of the National Academy of Sciences, Washington, D.C.

TABLE 22
SUGGESTED FORMAT FOR FORMULA CHARTS

Formula #: 19   Type of Ration: Corn Base Breeder   Calorie:Protein Ratio: 81.3:1

Ing. #	Ingredient	% in Mix	Protein %	Protein Amt.	Met. Energy kcal/lb	Met. Energy Amt.	Calcium %	Calcium Amt.	Total P %	Total P Amt.	Avail. P %	Avail. P Amt.	Vit. A IU/lb	Vit. A Amt.
5	Corn, yellow	.7075	.008	.06226	1559	1102.9925	.0002	.0001415	.0028	.001981	.0008	.000566	1318	932.485
22	Soybean meal, solv. 50%	.151	.508	.076708	1131	70.781	.0029	.0004379	.0065	.0009815	.002	.000302		
15	Meat and bone meal	.04	.504	.02016	891	35.64	.1010	.00404	.0496	.001984	.0496	.001984		
1	Alfalfa meal, dehy., 17%	.02	.175	.0035	623	12.46	.0144	.000288	.0022	.000044	.0022	.000044	83909	1678.18
10	DL-Methionine, 98%	.0005												
8	Dical. Phosphate, 18.5% P	.006					.21	.00126	.185	.00111	.185	.00111		
14	Limestone flour	.0625					.38	.02375						
20	Salt, iodized	.0025												
V/M	Vit. & Min. premix	.01											400000	4000.
	Totals	1.00		.162628		1321.8735		.0299174		.0061005		.004006		6610.665

Ing. #	Vit. D ICU/lb	Vit. D Amt.	Vit. E mg/lb	Vit. E Amt.	Vit. K mg/lb	Vit. K Amt.	Riboflavin mg/lb	Riboflavin Amt.	Total Niacin mg/lb	Total Niacin Amt.	Lysine %	Lysine Amt.	Methionine %	Methionine Amt.	Cystine %	Cystine Amt.
5			10.0	7.075			.46	.3219125	10.91	7.71175	.0024	.001698	.0020	.001415	.0015	.0010612
22			2.2	.3322			1.36	.20536	9.86	1.48886	.0319	.0048169	.0074	.0011174	.0083	.0012533
15			.4	.016			2.00	.08	20.91	.836	.0260	.00104	.0065	.00026	.0025	.0001
1			56.8	1.136	3.95	.079	6.18	.1236	17.27	.346	.0073	.000146	.0023	.000046	.0020	.00004
10													.98	.00049		
8																
14																
20																
V/M	50000	500	500	5.0	100	1.0	300	3.0	2500	25						
T		500		13.5592		1.079		3.7308725		35.38261		.007709		.0033284		.0024545

142

APPENDIX B

Symptoms of Vitamin and
Mineral Deficiencies in Ducks

Vitamin	Deficiency Symptoms*
VITAMIN A	Retarded growth; general weakness; staggering gait; ruffled plumage; low resistance to infections and internal parasites; eye infection; lowered production and fertility; increased mortality. (Adult birds develop symptoms slower than ducklings.) *Sources*: Fish-liver oils, yellow corn, alfalfa meal, fresh greens.
VITAMIN D₃	Retarded growth; rickets; birds walk as little as possible, and when they do move, their gait is unsteady and stiff; bills become soft and rubbery and are easily bent; thin-shelled eggs; reduced egg production; bones of wings and legs are fragile and easily broken; hatchability is lowered. *Sources*: Sunlight, fish-liver oils, synthetic sources.
VITAMIN E	Unsteady gait; ducklings suddenly become prostrated, lying with legs stretched out behind, head retracted over the back; head weaves from side to side; reduced hatchability of eggs; high mortality in newly hatched ducklings; sterility in males and reproductive failure in hens. *Sources*: Many feedstuffs both of plant and animal origin, particularly alfalfa meal, rice polish and bran, distiller's dried corn solubles, wheat middlings.

Summarized from *Nutrient Requirements of Poultry*, 7th revised edition (1977), pages 11-20, with the permission of the National Academy of Sciences, Washington, D.C.

* The deficiency symptoms are given in the sequence they normally evidence themselves.

Note: While work on vitamin and mineral deficiencies in poultry has been limited largely to chickens and turkeys, I have observed many of these symptoms in ducks.

VITAMIN K	Delayed clotting of blood; internal or external hemorrhaging which may result in birds bleeding to death from even small wounds. *Sources*: Fish meal, meat meal, alfalfa meal, fresh greens.
THIAMINE	Loss of appetite; sluggishness; emaciation; head tremors; convulsions; head retracted over the back. *Sources*: Grains and grain by-products.
RIBOFLAVIN	Diarrhea; retarded growth; curled-toe paralysis; drooping wings; birds fall back on their hocks; eggs hatch poorly. *Sources*: Dried yeast, skim milk, whey, alfalfa meal, green feeds.
NIACIN	Retarded growth; leg weakness; bowed legs; enlarged hocks; diarrhea; poor feather development. *Sources*: Dried yeast and synthetic sources. (Most of the niacin in cereal grains is unavailable to poultry.)
BIOTIN	Bottoms of feet are rough and calloused, with bleeding cracks; lesions develop in corners of mouth, spreading to area around the bill; eyelids eventually swell and stick shut; slipped tendon (see also choline and manganese deficiencies); eggs hatch poorly. *Sources*: Most feedstuffs, but especially dried yeast, whey, meat and bone meal, skim milk, alfalfa meal, soybean meal, green feeds. (The biotin in wheat and barley is mostly unavailable to poultry. If raw eggs are fed to animals, avidin—a protein in the egg-white—binds biotin, making it unavailable.)
PANTOTHENIC ACID	Retarded growth; viscous discharge causes eyelids to become granular and stick together; rough-looking feathers; scabs in corners of mouth and around vent; bottoms of feet rough and calloused, but lesions are seldom as severe as in a biotin deficiency; drop in egg production; reduced hatchability of eggs; poor liveability of newly hatched ducklings. *Sources*: All major feedstuffs, particularly brewer's and torula yeasts, whey, skim and buttermilk, fish solubles, wheat bran, alfalfa meal.
CHOLINE	Retarded growth; slipped tendons (see also biotin and manganese deficiencies). *Sources*: Most feedstuffs but especially fish meal, meat meal, soybean meal, cottonseed meal, wheat germ meal. (Evidence indicates that choline is synthesized by mature birds in quantities adequate for egg production.)

VITAMIN B$_6$ Poor appetites; extremely slow growth; nervousness; convulsions; jerky head movements; ducklings run about aimlessly, sometimes rolling over on their backs and rapidly paddling their feet; increased mortality.
Sources: Grains and seeds.

FOLACIN Retarded growth; poor feathering; colored feathers show a band of faded color; reduced egg production; decline in hatchability; occasionally slipped tendon.
Sources: Green feeds, fish meal, meat meal.

VITAMIN B$_{12}$ Poor hatchability; high mortality in newly hatched ducklings; retarded growth; poor feathering; degenerated gizzards; occasionally slipped tendon.
Sources: Animal products and synthetic sources.

Mineral **Deficiency Symptoms***

CALCIUM AND PHOSPHORUS Rickets; retarded growth; increased mortality; in rare cases, thin-shelled eggs.
Sources: Calcium—oyster shells and limestone. Phosphorus—dicalcium phosphate, soft rock phosphate, bone meal, meat and bone meal, fish meal. (Most all feedstuffs have varying amounts of calcium and phosphorus, but typically only about one-third of the phosphorus in plant products is available to birds.)

MAGNESIUM Ducklings go into brief convulsions and then lapse into a coma from which they usually recover if they are not swimming; rapid decline in egg production.
Sources: Most feedstuffs, especially limestone, meat and bone meals, grain brans. (Raising either the calcium or phosphorus content of feed magnifies a deficiency of this mineral.)

MANGANESE Slipped tendon in one or both legs; retarded growth; weak egg shells; reduced egg production and hatchability. Slipped tendon (also known as perosis) is first evidenced by the swelling and flattening of the hock joint, followed by the Achilles tendon slipping from its condyles (groove), causing the lower leg to project out to the side of the body at a severe angle. (Perosis can also be the result of biotin or choline deficiencies.)
Sources: Most feedstuffs, especially manganese sulfate, rice bran, limestone, oyster shell, wheat middlings and bran.

CHLORIDE
Extremely slow growth; high mortality; unnatural nervousness.
Sources: Most feedstuffs, especially animal products, beet molasses, alfalfa meal.

COPPER
AND IRON
Anemia.
Sources: Most feedstuffs, including fresh greens.

IODINE
Goiter (enlargement of the thyroid gland); decreased hatchability of eggs.
Sources: Iodine. Iodized salt contains such small quantities of iodine that it cannot be relied upon to provide sufficient iodine.

POTASSIUM
Rare, but when it occurs, results in retarded growth and high mortality.
Sources: Most feedstuffs.

SODIUM
Poor growth; cannibalism; decreased egg production.
Sources: Most feedstuffs, but especially salt and animal products.

ZINC
Retarded growth; moderately to severely frayed feathers; enlarged hock joints; slipped tendon.
Sources: Zinc oxide and most feedstuffs, especially animal products.

Predators

One of the most frustrating experiences for the owner of ducks is to lose valuable birds and eggs to predators. The seriousness of this problem varies a great deal from one location to another. It is a good idea to talk with poultry keepers in your area to learn what kind of problems they have encountered and what precautions are necessary to minimize losses.

There are various species of wild and domestic animals that will prey upon ducks. Of the domestic type, dogs—and to a lesser extent, cats—are notorious for the damage they can inflict on poultry flocks. The first time one of these pets shows an interest in your ducks, they must be disciplined *immediately* if future troubles are to be avoided.

Some of the wild creatures that cause duck raisers grief are rats, ground squirrels, weasels, mink, skunks, raccoons, opossum, foxes, coyotes, turtles, snakes, crows, ravens, magpies, jays, gulls, hawks, falcons and owls. Because these animals—with the possible exception of rats—are essential in maintaining the delicate balance of nature, we must not attempt to eliminate them, but rather, need to respect them and give poultry sufficient protection so that hungry predators passing by will not be able to dine at the expense of our birds.

SECURITY MEASURES

Ducks roost on the ground, making them most vulnerable at night. Because many predators have nocturnal habits, a small building or fenced yard where ducks can be locked in after dark is a *must* in most localities. If you try to raise ducks without penning them up nightly, I can predict with almost 100 percent accuracy that your flock will be ravaged sooner or later.

A sturdy woven wire fence at least four feet high goes a long way in keeping ducks safe while they sleep. In areas where determined hunters such as raccoons and opossum are present, electric fencing can be used in combination with woven wire fences to make yards safer after sundown. To be the most effective, two strands of electric wire need to be utilized. One strand should run around the outside of the woven wire four inches above the ground, while the second strand should be installed a couple of inches above the top of the fence.

If weasels, mink, owls and cats (wild or domestic) are prevalent in your locality, at nighttime it will be expedient to pen your ducks in a shelter having a covered top as well as sturdy sides. Some predators will dig under fences or dirt floors of houses to gain entrance to impenned birds. Burying the bottom six to twelve inches of fences that encircle yards or covering the floors of duckhouses with wire netting will keep out excavators.

Because setting hens and ducklings are especially vulnerable to predation, extra care must be taken to provide them with secure quarters. Setting hens should be encouraged to nest in shelters that can be closed at night. When hens do nest in the open, panels four feet high can be set up around them to provide protection.

IDENTIFYING THE CULPRIT

If you do lose birds or eggs, it is helpful to be able to identify the culprit in order to remedy the situation. Once predators discover a convenient source of food, they often return at *regular* intervals if permitted to do so. The following is a brief guide to help you recognize the work of some of the more common predators, with suggestions on how to stop their thievery.

Dogs

Telltale signs. A number of ducks, sometimes the entire flock, badly maimed. Check for large holes under or through fences, and clumps of dog hair caught on wire.

Stopping losses. Strong fences at least four feet high.

Cats (wild or tame)

Telltale signs. Crushed eggs held together by shell membranes. Birds disappear without a trace, or only a few feathers or clumps of down are found in a secluded spot where the animal fed.

Stopping losses. Keep ducklings in wire-covered runs until they are two to four weeks old. Domestic cats are intelligent enough that they can be taught fairly easily not to molest poultry. When a pet cat is seen stalking your birds, let the tabby know *immediately* that the ducks are off limits. Throwing a rolled up newspaper at the offending animal will usually get the message across. If you have problems with stray cats, a live trap may be needed.

Foxes

Telltale signs. Foxes are fastidious hunters and normally leave little evidence of their visits. Usually they kill just one duck at a time and take the bird with them or partially bury it nearby. (Foxes have been known to go on rampages, killing thirty or more birds at a time and scattering carcasses over a quarter-mile area.) You might be able to find a small hole under or through fences, or a poorly fitted gate or door pushed ajar.

Stopping losses. Tight fencing at least four feet high with two strands of barbed or electric wire, and close-fitting gates. Foxes will squeeze or dig under fences that are not flush with the ground.

Raccoons

Telltale signs. End of eggs bitten off, or crops (esophagus) eaten out of dead birds—possibly heads missing. Usually returns every fourth or fifth night.

Stopping losses. Fences four feet high with two strands of electric wire, or lock birds at night in shelter with covered top and sides.

Skunks

Telltale signs. Destroyed nest with crushed shells mixed with nest debris.

Stopping losses. Gather eggs daily. Encourage setting hens to nest in shelters that can be locked up at night, or set up panels around hens that are nesting in the open.

Opposum

Telltale signs. Smashed eggs and birds that are badly mauled.

Stopping losses. At night, lock birds in a shelter with covered top and sides.

Mink and Weasels

Telltale signs. Young ducklings disappear or larger birds killed, evidently for amusement, with small teeth marks on head and neck.

Stopping losses. At nighttime, lock the birds in a shelter that is covered with ½-inch wire hardware cloth. As incredible as it seems, mink and weasels can pass through holes as small as 1 inch in diameter.

Snapping Turtles and Large Fish

Telltale signs. Ducklings disappear mysteriously while swimming.

Stopping losses. If your ducks frequent bodies of water that host turtles or large fish such as northern pike and large mouth bass, keep ducklings away from the water until they are two to four weeks old.

Snakes

Telltale signs. Nest appears untouched, but some or all eggs are missing.

Stopping losses. Encourage setting hens to lay in covered nest boxes.

Rats

Telltale signs. Eggs or dead ducklings pulled into underground tunnels.

Stopping losses. Until ducklings are three to six weeks old, at night put them in a pen that has sides, top and floor that are covered with ½-inch wire mesh. Rat populations should be kept under control with cats or traps.

Crows, Jays, Magpies and Gulls

Telltale signs. Punctured eggs or shells scattered around base of an elevated perch such as a fencepost or tree stump. These birds also occasionally steal newly hatched ducklings.

Stopping losses. Provide covered nest boxes and gather eggs several times a day. Keep ducklings in wire-covered runs until they are at least two weeks old.

Hawks

Telltale signs. Ducklings disappear during daylight without a trace or only a few scattered feathers or clumps of down.

Hawks are probably falsely accused of stealing poultry more often than any other predator. Because the large, soaring hawks (buteos) are so visible, people mistakenly assume these winged hunters are the cause of every missing barnyard fowl. Actually, the hawks we need to worry about are the ones that are seldom seen by the casual observer; the accipiter family which includes the Goshawk, Cooper's Hawk and Sharp-shinned Hawk. The accipiters are secretive—sticking to trees—and hunt from low altitudes. They can be identified by their short, round wings and long tails. If you do occasionally lose ducklings to hawks, your frustration may be reduced if you'll remember that the average hawk eats 200 to 300 rodents yearly.

Stopping losses. Keep ducklings in wire-covered runs until they are two to four weeks old.

Owls

Telltale signs. One or more ducks killed nightly with head and neck eaten.

Stopping losses. At night, lock ducklings and adult ducks in a shelter covered with 1" × 1" wire netting. Some of the smaller species (such as Screech Owls) can squeeze through 2" × 2" wire mesh.

Storing Eating Eggs

Several methods can be used for short or long term storage of eating eggs. Whichever mode you use, keep in mind that the quality of eggs decreases *rapidly* when they are exposed to temperatures above 60° F., sunlight, dry air or contamination due to soiled or cracked shells. Infertile eggs can be stored longer than fertile ones.

Dirty eggs should be washed with warm water soon after being gathered, and used within one or two weeks. Warm water opens the shell pores, drawing out filth that would remain embedded if a cold bath were used.

NATURAL STORAGE

Nest-clean, strong-shelled duck eggs can be held for two or three weeks without refrigeration if they are placed in a cool (below 60° F.) humid, dark nook. Cellars, pumphouses, garages and unheated basements often meet these requirements.

REFRIGERATION

Under refrigeration (34° to 40° F.), eggs can be kept safely for up to six weeks. By sealing freshly laid eggs in plastic bags, their refrigeration life can be lengthened to two months.

FREEZING

Freezing is an excellent way to save surplus eggs for seasons when production slacks off. By following strict sanitation precautions in preparing waterfowl eggs for the freezer, we have been able to keep them for over twelve months successfully. However, because frozen eggs of any type can harbor bacteria that may cause acute intestinal infections, it is safest to store them *no more* than six months and restrict their use for baked or long-cooked foods.

Select only fresh, nest-clean eggs for freezing. Wash them with warm water, and then break into a boiled mixing bowl. Blend the whites and yolks with a fork, being careful not to beat in air bubbles. To prevent the yolks from becoming excessively thick during storage, add either one teaspoon of salt or one tablespoon of honey to each pint of eggs.

The last step is to pour the batter into thoroughly clean ice cube trays or freezer containers, leaving ½ inch headroom to allow for expansion. If ice trays are used, remove the egg cubes soon after they are frozen solid, seal them in a clean container and place immediately in the freezer.

Once frozen eggs are thawed, they should be used within twenty-four hours. *Do not* refreeze thawed eggs.

Duck Recipes

There are many ways duck can be fixed to taste great. Unfortunately, duck cookery is ignored or passed over lightly in most cookbooks. So with a lot of help from Millie, I'd like to share a few of our favorite recipes.

FRIED DUCKLING

1 duckling, cut up into pieces
   flour
   salt
   pepper

Dust pieces of duckling with flour. Season with salt and pepper. Fry in lightly greased skillet over medium-low heat for 1½ hours. Turn pieces as necessary to brown evenly.

Remove fat from skillet as it fries out of duck. Use drippings for gravy. Serves 4 to 6.

ROAST DUCKLING

1 whole 4½-5 lb. duckling
   salt
   pepper
   stuffing

Rub inside of duck cavity with salt. The duck may be roasted with or without stuffing. Fasten opening with skewers and lace closed with strong thread or thin string.

Place duckling uncovered on rack of roasting pan. Roast in preheated oven at 325° F. for 2½ to 3 hours, until skin is crisp and brown and flesh is tender.

Serves 4 to 6.

STEAM-FRIED DUCK EGGS

Put eggs into medium-hot, lightly greased skillet. Pour a small amount of hot water around eggs and place lid on skillet. Steam-fry for several minutes until egg-white is set. Serve immediately.

DUCK STEW

In a large kettle:

Sauté: 1-2 cloves garlic, minced
              2 tbsp. butter or margarine
Add: 1-2 cups leftover duck meat, in chunks
              3 medium-sized potatoes, diced
              4 carrots, sliced in ½-inch pieces
              1 large onion, wedged
              1 cup shredded cabbage
              2 cups tomatoes

Cover with vegetable or meat stock. Season with salt, pepper, paprika and desired herbs. Cook over medium-low heat till vegetables are crispy tender. Add water if necessary to keep vegetables covered.

Add: 1 cup peas, green beans or limas
      2-3 celery stalks, in chunks
        1 cup corn

Heat to boiling point (or longer if fresh or frozen beans are used). If you prefer a thicker stew, add a paste of flour and water. Serve with fresh, warm homemade bread.
Serves 6 to 8.

BARBECUED DUCKLING

    1 whole 4-5 lb. duckling
      salt
      pepper

Have plenty of hot coals ready. Season inside cavity of duckling. Fasten bird securely to rotisserie rod (or pole, if you're cooking it over an open pit). Salt and pepper outside of duckling. Position bird 6 to 12 inches from coals. Relax and enjoy the aroma as the duck sizzles and cooks for 2½ to 3 hours.

When done, serve piping hot with sauerkraut, mashed potatoes, and homemade bread.
Serves 4 to 6.

USING GIBLETS

Livers: We like them made into patties. Grind up (or blend) raw livers, along with onion, green pepper and a slice of bacon for extra flavor. Add an egg, a little flour; season with salt and pepper. Spoon into lightly greased skillet and fry on both sides till brown.

Hearts and gizzards: Bake them while you're using the oven anyway. Slice and use on pizza—they're delicious! Use hearts and gizzards to make gravy that tastes yummy over pancakes or on mashed potatoes.

STIR-FRIED VEGETABLES AND DUCK

Combine in small bowl and set aside:

1-2 tbsp. cornstarch
1 tsp. salt
1 tbsp. honey
3 tbsp. soy sauce
1½ cups stock or water

Cut raw vegetables (enough to make 5 cups) such as the following, in any combination: asparagus, broccoli, cabbage, carrots, cauliflower, green beans or zucchini. Set vegetables aside, grouping according to cooking time.
Heat in skillet:

4 tbsp. cooking oil
Add: 2-3 cloves garlic, minced
1 large onion, sliced in thin wedges
½-1 cup leftover duck, cut in strips or chunks

Stir-fry over medium-high heat for 3 to 5 minutes. Add longest cooking vegetables, such as cabbage, carrots and green beans, continuing to stir-fry. As soon as they begin to tenderize, add faster cooking vegetables. If necessary, add more oil. When vegetables are crispy-tender—and still have their bright color—add soy sauce mixture. Cook, while stirring, till sauce is clear. Serve *immediately* over hot rice.
Serves 4 to 6.
Note: Have all other meal preparations finished when you begin with this part—it takes a lot of last-minute work!

HOMEMADE NOODLES

¾ cup whole wheat flour
¾ cup unbleached white flour
1 tsp. salt
1 tsp. fat or margarine
4 duck egg yolks (or 2 whole eggs), slightly beaten

Combine flours and salt. Rub fat into dry ingredients to form coarse crumbs. Add eggs to form a stiff dough. Knead 25 strokes on floured board. Divide dough into three parts and roll each as thin as possible.
Then, if you have a hand noodle slicer, cut at this point, then dry strips in warm oven.
Otherwise, spread rolled dough on a cloth and allow to partially dry; then roll dough as a jelly roll and cut thinly.
Use immediately or store in refrigerator or freezer.
Makes approximately 12 ounces.

DEL'S DELIGHTFUL OMELET

4 duck eggs
pepper
2 tbsp. butter or margarine
chopped vegetables, such as onion, zucchini, tomato, egg-
plant, green or red pepper, mushrooms, broccoli
cheddar cheese, shredded alfalfa sprouts

Beat eggs. Melt butter in skillet and add egg mixture. Add your choice of vegetables and season with pepper.

Cook slowly till underside is *barely* brown. Flip, then sprinkle cheese on top. When egg has set and cheese has melted, serve with sprouts.

Serves 4 to 6.

WHOLE WHEAT ANGEL FOOD CAKE

Mix: 1 cup sifted whole wheat flour
         ¾ cup packed brown sugar

Crush and mix sugar lumps.
Beat together until stiff but glossy:

         1½ cups duck egg-whites (8-10 eggs)
         1½ tsp. cream of tartar
         ¼ tsp. salt
         1 tsp. vanilla
         ½ tsp. almond flavoring

Add: ¾ cup packed brown sugar to beaten whites, ¼ cup at a
         time, beating well after each addition.

Fold in flour with a large spoon, sifting a little over the top, folding in lightly with a down-up-over motion. When well-blended, pour into an ungreased 10-inch angel food cake pan. Bake at 375° for 45 to 60 minutes; touch the top gently to see if cake is done.

Invert pan and cool cake thoroughly. Remove from pan.

Option: Substitute ¼ cup carob powder for an equal amount of flour and omit almond flavoring.

Using Feathers and Down

Those soft feathers and down that were carefully saved at butchering time can be used in making a variety of handy items. And as the prices of manufactured down-filled articles continue to spiral upward, home-grown feathers are becoming more practical all the time.

While all ducks produce good-quality plumage, the down of Muscovies is less desirable than that of the breeds derived from Mallards. Large breeds such as Aylesbury, Pekin and Rouen will produce 2½ to 3½ ounces of down and small feathers per bird. Mature ducks have more down than ducklings that are seven to twelve weeks old, and birds have more down in the late fall after cold weather has set in than during the warm months of summer.

PLUCKING LIVE DUCKS

People frequently ask about plucking live ducks for down. While ducks usually survive such treatment, this practice places a great deal of stress on the birds and reduces their productivity. If ducks are live-plucked, taking the following precautions will reduce the negative effects it has on the birds.

1. Pick the birds only once a year during the late spring or early summer after they have begun their natural molt. At this time, their feathers will be loosened in the follicles, and the plumes can be removed with less pain to the birds.

2. Pull out only small pinches of feathers and down at a time to keep from tearing the bird's skin.

3. Take feathers only from the underside of the duck.

4. Remove a maximum of 50 percent of the feathers from the plucked area, and do not leave any bare patches.

5. Do not let the birds swim for two or three weeks, or until they have grown new feathers.

PILLOWS

A tightly woven cloth, such as down-proof ticking, is essential to keep down and feathers from working their way out of the pillow. Double stitched seams should be used to keep the exits closed.

Small, soft body feathers need to be mixed with down to make pillows more resilient. A pillow filled with just down is too weak to hold its shape well. An excellent ratio is 75 percent down and 25 percent feathers, although good pillows can be obtained with a mixture of half down and half feathers.

COMFORTERS, QUILTS AND SLEEPING BAGS

A down-filled coverlet—what a way to keep warm on cold winter nights! Not only are they delightful, but also efficient. Millions of people around the globe are still being warmed by feathers as they sleep.

The tops and bottoms of comforters, quilts or sleeping bags must be lined with down-proof material. To keep the down and feathers distributed evenly, channels five to six inches wide need to be made. Leave one end of the channels open for stuffing.

If the large plumes of the wings, tail and body have been discarded at picking time, feathers and down can be used in the same ratio they come off the bird. As each channel is filled, sew the opening shut by hand; then finish the edges with binding.

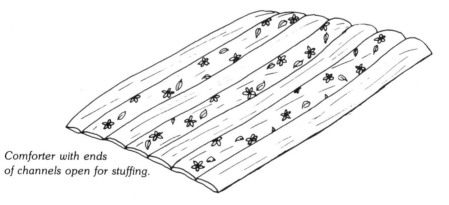

Comforter with ends of channels open for stuffing.

CLOTHING

As in comforters, you will need to make channels to keep the warmth spread around. For parkas, vests, etc., make the channels narrower, approximately two to three inches wide. Remember to use down-proof lining, and to make those seams double so that none of the down escapes. To keep clothing as lightweight as possible, use 75 to 90 percent down, with small quantities of body feathers.

APPENDIX G

Duck Breeders and Hatchery Guide

The Duck Breeders and Hatchery Guide was prepared to help you locate sources of ducks. This list is not exhaustive. Please keep in mind that the information is dated, and details such as addresses, stock sold, and so on may change with time.

While attempts have been made to include only reliable sources, I am *not* endorsing the quality of stock or service given by the following firms. Some hatcheries sell crossbred ducklings under purebred labels. If you are interested in acquiring a specific breed, ask potential sources if they *guarantee* their stock to be pure-blooded.

Three additional sources for addresses of duck breeders are *Feather Fancier* and *Poultry Press* (see Appendix I, page 166, for addresses) and the breeders directory of the Society for Preservation of Poultry Antiquities. The directory is available from the society's secretary (James Rice, Route 3, Greenwood, Wisconsin 54437) for $5 per copy.

When writing to poultry breeders and hatcheries, always have the courtesy to enclose a stamp. Some hatcheries and brokers will send their catalogues only if you send a dollar (which is refunded with your first order) along with a request for particulars on the stock they carry.

Code: Hatcheries with no asterisk offer production-bred stock, while those with one asterisk sell standard-bred ducks. Two asterisks indicate that both production-bred and standard-bred birds are available.

California

Blohm's, 54 Blanca Lane, Watsonville, CA 95076—Ph. (408) 724-6743
Ducklings in Khaki Campbell; Pekin; Rouen.

Hockman's, Box 7187, San Diego, CA 92107—Ph. (714) 222-6983
Ducklings in Khaki Campbell; White Muscovy; Pekin; Rouen; Mallard.

Metzer Hatchery, 25870 Old Stage Rd., Gonzales, CA 93926—Ph. (408) 679-2355
Ducklings in Khaki Campbell; Fawn & White Runner; White Muscovy; Pekin; Rouen; Cayuga; Buff Orpington; Swedish; Mallard.

Connecticut

*Charles Herrman, 610 Center St., Manchester, CT 06040
Hatching eggs and ducklings in Khaki Campbell; White, Fawn & White Runner; White Muscovy; Pekin; White Crested; Swedish.

Florida

**John Thiele, Box 1545, Hobe Sound, FL 33455—Ph. (305) 546-4675
Ducklings and mature stock in Khaki Campbell; White, Fawn & White, Black Runner; Pekin; Rouen; Cayuga; White Crested; Buff Orpington; Swedish; White, Gray Call; East Indie.

Morris Hatchery, 8900 S.W. 80th St., Miami, FL 33173—Ph. (305) 271-8982
Ducklings in Pekin; Rouen; Mallard.

Illinois

*Stephen F. Gerdes, Rt. 1, Toluca, IL 61369—Ph. (815) 452-2534
Hatching eggs, ducklings and mature stock in Khaki Campbell; White, Fawn & White, Penciled, Solid Fawn, Blue, Black, Chocolate, Buff, Gray Runner; Aylesbury; White, Colored, Blue, Chocolate, Silver, Buff Muscovy; Pekin; Rouen; Cayuga; White, Gray, Black Crested; Magpie; Buff Orpington; Swedish; Australian Spotted; White, Gray Call; East Indie; Gray, White, Snowy, Golden Mallard.

Louisiana

*Clark's Bayou Bird Farm, Star Rt., Box. 192½, Princeton, LA 71067—
Ph. (318) 949-2294
Mature stock in White, Black Runner; Blue Muscovy; White Crested; Swedish; East Indie; Gray, White, Golden, Crested Mallard.

Massachusetts

*Roger A. Sanford, 832 Pine Hill Rd., South Westport, MA 02790—Ph. (617) 636-2533
Hatching eggs and ducklings in White Runner; Pekin; Rouen; White, Gray Call; East Indie.

Minnesota

Leuze's Waterfowl Hatchery, Rt. 5, Willmar, MN 56201
Ducklings in Rouen.

**Neubert Hatcheries, P.O. Box 1239, Mankato, MN 56001
Hatching eggs, ducklings and mature stock in Runner; White, Colored Muscovy; Pekin; Rouen; Crested; Gray, White Mallard.

Pietrus Hatchery, 112 E. Pine St., Sleepy Eye, MN 56085—Ph. (507) 794-3411
Ducklings in White Muscovy; Pekin; Rouen; White Mallard.

Sobania Hatchery, Rt. 6, Little Falls, MN 56345
Ducklings in Khaki Campbell; Pekin; Rouen; White, Gray Crested; Buff Orpington; Gray, White Mallard.

**Stromberg's, Fifty Lakes Rt., Pine River, MN 56474
Hatching eggs, ducklings and mature stock in Khaki Campbell; Runner; Muscovy; Pekin; Rouen; Cayuga; Crested; Buff Orpington; Swedish; Mallard.

**Sunny Creek Farms & Hatchery, Rt. 2, Red Lake Falls, MN 56750—
Ph. (218) 253-2211
Hatching eggs and ducklings in Khaki Campbell; White, Fawn & White, Penciled Runner;
Aylesbury; White, Colored, Blue Muscovy; Pekin; Rouen; Cayuga; White Crested; Buff
Orpington; Swedish; White, Gray Call; East Indie; Mallard.

*Duane Urch, Rt. 1, Box 48, Owatonna, MN 55060—Ph. (507) 451-6782
Hatching eggs, ducklings and mature stock in Khaki Campbell; White, Fawn & White,
Penciled Runner; Aylesbury; White, Colored, Blue Muscovy; Pekin; Rouen; Cayuga;
White Crested; Buff Orpington; Swedish; East Indie.

Missouri

Cackle Hatchery, P.O. Box 529, Labanon, MO 65536—Ph. (417) 532-4581
Ducklings in Khaki Campbell; Pekin; Rouen.

Crow Poultry & Supply, Box 339, Windsor, MO 65360
Ducklings in Pekin; Rouen.

Heart of Missouri Poultry Farm, P.O. Box 954, Columbia, MO 65201—
Ph. (314) 442-3534
Ducklings in White Muscovy; Pekin; Rouen; Mallard.

*Zillich Poultry Farm, Rt. 1, Mercer, MO 64661—Ph. (816) 875-2572
Hatching eggs and ducklings in Penciled, Black, Blue, Buff Runner; Pekin; Cayuga;
Swedish.

North Dakota

Magic City Hatchery, P.O. Box 1771, Minot, ND 58701—Ph. (701) 839-3726
Ducklings in Pekin; Rouen.

Ohio

Pilgrim Goose Hatchery, Creek Rd., Williamsfield, OH 44093
Ducklings in Runner; Aylesbury; Pekin; Rouen; Cayuga; Buff Orpington; Swedish.

Pruden Hatchery, Rt. 84 & Meyers Rd., Geneva, OH 44041—Ph. (216) 466-1773
Ducklings in Pekin; Rouen; Mallard.

Ridgway Hatcheries, Inc., LaRue, OH 43332
Ducklings in Pekin; Rouen; Mallard.

**Ronson Farms, P.O. Box 12565, Columbus, OH 43212—Ph. (614) 486-6219
Ducklings and mature stock in Khaki Campbell; White, Fawn & White, Penciled, Gray
Runner; White, Colored Muscovy; Pekin; Rouen; Cayuga; White Crested; Buff Orpington;
Swedish; White, Gray Call; East Indie; Mallard.

Oklahoma

Country Hatchery, Box 747, Wewoka, OK 74884—Ph. (405) 257-3250
Ducklings in Pekin; Buff Orpington; Swedish; Mallard.

Oregon

Mother Hen Farm & Hatchery, 3810 NE Hwy. 20, Corvallis, OR 97330—Ph. (503)
752-6520
Hatching eggs and ducklings in White and Khaki Campbell; White Runner; Mallard.

**Northwest Farms, P.O. Box 3003, Portland, OR 97208—Ph. (503) 653-0344
Hatching eggs, ducklings and mature stock in Khaki Campbell; White, Fawn & White, Penciled Runner; Aylesbury; White, Colored, Blue Muscovy; Pekin; Rouen; Cayuga; White, Gray Crested; Buff Orpington; Swedish; White, Gray Call; East Indie; Mallard.

Shank's Hatchery, P.O. Box 429, Hubbard, OR 97032—Ph. (503) 981-7801
Ducklings in Pekin; Mallard.

Pennsylvania

Clearview Hatchery, Gratz, PA 17030—Ph. (717) 365-3234
Ducklings in Khaki Campbell; White Muscovy; Pekin; Rouen; Cayuga; Crested; Buff Orpington; Swedish; White Call; East Indie; Mallard.

*Feather Edge Farm, Rt. 1, Cochranton, PA 16314
Hatching eggs, ducklings and mature stock in Khaki Campbell; Pekin; Cayuga; White, Black Crested; Buff Orpington; White Call.

Hoffman Hatchery, Gratz, PA 17030—Ph. (717) 365-3407
Ducklings in Khaki Campbell; Fawn & White Runner; White Muscovy; Pekin; Rouen; Cayuga; Buff Orpington; Swedish; Mallard.

**Willow Hill Farm & Hatchery, Rt. 1, Box 100, Richland, PA 17087—
Ph. (717) 933-4606
Ducklings and mature stock in Khaki Campbell; White, Fawn & White, Penciled Runner; White Muscovy; Pekin; Rouen; Cayuga; White, Gray, Black, Blue Crested; Buff Orpington; Swedish; White, Gray Call; East Indie; Mallard.

South Dakota

Inman Hatchery, 3000 Third Ave., S.E., Aberdeen, SD 57401—Ph. (605) 225-8122
Ducklings in Pekin; Rouen.

Washington

Harder's Hatchery, Rt. 101, Box 316, Ritzville, WA 99169—Ph. (509) 659-1423
Ducklings in Khaki Campbell; Pekin; Rouen.

Wisconsin

Abendroth's Hatchery, Rt. 2, Box 200, Waterloo, WI 53594—Ph. (414) 478-2053
Ducklings in Khaki Campbell; Fawn & White, Penciled Runner; White Muscovy; Pekin; Rouen; Cayuga; White Crested; Buff Orpington; Swedish; Gray, White Mallard.

*Halbach Poultry Farm, 305 S. Third St., Waterford, WI 53185—
Ph. (414) 534-6405
Hatching eggs, ducklings and mature stock in Khaki Campbell; White, Fawn & White, Penciled, Black, Gray Runner; Aylesbury; Pekin; Rouen; Cayuga; White Crested; Buff Orpington; Swedish; White, Gray Call; East Indie.

Canada

Berg's Hatchery, Box 603, Russel, Manitoba ROJ 1WO
Ducklings in Pekin; Rouen.

Canadian Poultry Supplies, Rt. 2, Lindsay, Ontario K9V 4R2
Hatching eggs, ducklings and mature stock in Khaki Campbell; White, Fawn & White, Penciled, Black Runner; Aylesbury; White, Colored, Blue Muscovy; Pekin; Rouen; Cayuga; White, Black Crested; Black, Blue Magpie; Buff Orpington; Swedish; White, Gray Call; East Indie; Mallard.

Miller Hatcheries Ltd., 260 Main St., Winnipeg, Manitoba R3C 1A9—
Ph. 943-6541
(Branch offices in Regina, Saskatoon, North Battlefield, Edmonton)
Ducklings in Pekin; Rouen.

Springhill Hatchery, Neepawa, Manitoba ROJ 1HO
Ducklings in Pekin.

Pardo's Hatchery, Rt. 3, Blenheim, Ontario
Ducklings in Pekin; Rouen.

Webfoot Farm & Hatchery Ltd., Elora, Ontario NOB 1SO—Ph. (519) 846-9885
Ducklings in White Muscovy; Pekin; Rouen.

Puerto Rico

Aibonito Hatchery, Ruta 2, Buzon 490, Aibonito, PR 00609—Ph. (809) 735-8585
Ducklings in several breeds.

Sources of Supplies and Equipment

One of the advantages of ducks is that a minimum of special equipment and supplies is required in order to raise them successfully. However, there may be times when you'll need an incubator, a few bands, some picking wax, etc., but cannot find a local source for these products. For this reason, the following firms are listed.

COMPLETE INVENTORY

Burkey Co.
P.O. Box 29465
San Antonio, TX 78229
Ph. (512) 696-0706

Canadian Poultry Supplies
Rt. 2
Lindsay, Ontario
Canada K9V 4R2

College Poultry Supplies
287 College St.
Toronto, Ontario
Canada M5T 1S2
Ph. (416) 924-5598

Foy's Supplies
Box 27166
Golden Valley, MN 55427

Gragne Bros. Supplies
2883 Woodland Circle
Allison Park, PA 15101
Ph. (412) 443-2486

Marsh Manufacturing, Inc.
14232 Brookhurst St.
Garden Grove, CA 92643
Ph. (714) 534-6580

Northwest Farms, Inc.
P.O. Box 3003
Portland, OR 97208
Ph. (503) 653-0344

Rocky Top Poultry Supply
P.O. Box 1006
Harriman, TN 37748
Ph. (615) 882-8867

Ronson Farms
P.O. Box 12565
Columbus, OH 43212
Ph. (614) 486-6219

Sidney Shoemaker Poultry Supplies
3091 Lincoln-Gilead Twp. Rd. 124
Rt. 3
Cardington, OH 43315
Ph. (419) 864-6666

Stocklin Supply Co.
738 S.E. Lincoln
Portland, OR 98124
Ph. (503) 234-0897

Strecker's Poultry Supply
Rt. 3, Box 365-K
Arroyo Grande, CA 93420

Stromberg's
50 Lakes Route
Pine River, MN 56474
Ph. (218) 543-4223

Valentine Equipment Co.
9706 S. Industrial Drive
Bridgeview, IL 60455
Ph. (312) 599-1101

INCUBATORS AND BROODING EQUIPMENT ONLY

Brower Manufacturing Co.
Box 5722
Quincy, Il 62301

Hockmans
12659 Devonshire
San Diego, CA 92107
Ph. (714) 222-6983

Leahy Manufacturing Co.
406 W. 22nd St.
Higginsville, MO 64037
Ph. (816) 584-2641

The Humidaire Incubator Co.
217 W. Wyne St.
New Madison, OH 45346
Ph. (513) 996-3001

Oak Ridge Manufacturing Co.
Veyo, UT 84778
Ph. (801) 673-9190

Superior Incubator Co.
4734 Sanford
Houston, TX 77035
Ph. (713) 729-2109

Suggested Reading

MAGAZINES AND NEWSPAPERS

Backyard Poultry, Jerome D. Belanger, publisher, Rt. 1, Waterloo, Wisconsin 53594. $7 yearly. A monthly publication devoted to the small flock owner, which carries articles on most types of poultry.

Countryside, Jerome D. Belanger, editor and publisher, Rt. 1, Waterloo, Wisconsin 53594. $9 yearly. Authoritative monthly magazine for small farmers. Regular features on poultry with question and answer department.

Feather Fancier, Corey R. Herrington, editor and publisher, P.O. Box 239, Erin, Ontario, Canada NOB 1TO. $6 yearly, 50¢ for sample. Monthly paper devoted to standard-bred poultry, pigeons and pet stock. Good source for addresses of Canadian poultry breeders and hatcheries.

Poultry Press, Robert F. DeLancey, editor and publisher, Box 947, York, Pennsylvania 17405. $5 yearly, 50¢ for sample. Monthly paper with articles on poultry shows, club news, and breeding and management tips. Good source for addresses of waterfowl breeders and hatcheries throughout North America.

BOOKS

Light Weight Camping Equipment and How to Make It, by Gerry Cunningham and Margaret Hansson. Available from Gerry Division of Outdoor Sports Industries, 5450 North Valley Hwy., Denver, Colorado 80216. Gives patterns and instructions for making down-filled clothing.

Nutrient Requirements of Poultry, 7th revised edition (1977), National Academy of Sciences, 2101 Constitution Avenue, Washington, D.C. 20418. An informative though brief manual on poultry nutrition. Highly recommended for persons who desire a working understanding of poultry nutrition and who plan to formulate rations.

Poultry Feed Formulas, Department of Animal and Poultry Science, Ontario Agricultural College, University of Guelph, Guelph, Ontario, Canada. Contains information on ration formulation for most types of poultry and includes 28 duck rations using various feedstuffs.

Modern Waterfowl Management and Breeding Guide, by Oscar Grow. Available from American Bantam Association, P.O. Box 610, N. Amherst, Massachusetts 01059. $12. Contains detailed information on the origin, history and breeding of most breeds and species of domestic and semi-domestic ducks, geese and swans. 359 pages, numerous illustrations, hardback.

Successful Duck & Goose Raising, by Darrel Sheraw. Available from Stromberg Publishing Co., Pine River, Minnesota 56474. $5.95. Most comprehensive book available on selecting and mating standard-bred waterfowl. 208 pages, 225 pictures, paperback.

Standard of Perfection for Domesticated Land Fowl and Water Fowl, American Poultry Association, Inc., Box 70, Cushing, Oklahoma 74023. $12. Describes in detail all varieties of large chickens, bantams, turkeys, ducks and geese recognized by the American Poultry Association. Of special interest for persons who raise pure-bred poultry. 600 pages, over 200 illustrations, hardback.

Organizations

International Waterfowl Breeders Association; Lyle Jones, Secy., 12402 Curtis Rd., Grass Lake, MI 49240. Dues $3. Sponsors meets at shows and sends out newsletters (2-4 pages).

Associated Breeders of Campbell Ducks; Rick Luttman, Secy., 7909 Lynch Rd., Sebastopol, CA 95472. Dues $3. An active organization that sends out quarterly newsletters (6-8 pages) containing information on all aspects of breeding and managing Campbells for exhibition and egg production.

American Poultry Association, Inc.; Allen D. Fitchett, Exec. Secy., Box 70, Cushing, OK 74023. Dues $4. For breeders of all poultry. Membership includes newsletter and yearbook (112 pages in 1977) containing articles and advertisements from leading poultry breeders.

American Bantam Association; Fred P. Jeffery, Secy., P.O. Box 610, N. Amherst, MA 01059. Dues $5. For breeders of bantam chickens and ducks. All members receive the A.B.A. quarterly newsletter, and new members are sent a free copy of the A.B.A. yearbook (212 pages in 1977).

Society for Preservation of Poultry Antiquities; James K. Rice, Secy-Treas., Rt. 3, Greenwood, WI 54437. Dues $5. Members receive S.P.P.A. Breeder Directory (last issue contained 97 pages and listed 184 poultry breeders throughout North America) and quarterly newsletters (6-12 pages) with information on all types of rare poultry.

Bibliography

American Poultry Association, Inc. *The American Standard of Perfection.* Cushing, Oklahoma: 1973.

Bernsohn, Ken. "Producing Your Own Down." *Organic Gardening and Farming,* Vol. 25, No. 3 (March 1978), p. 114.

Chatterton, F.J.S. *Ducks and Geese and How to Keep Them.* London: 1924.

Cornell University Agricultural Experiment Station. Department of Poultry Science. *Duck Rations.* Extension Stencil No. 25, 1973.

Grow, Oscar. *Modern Waterfowl Management and Breeding Guide.* American Bantam Association, 1972.

Ives, Paul. *Domestic Geese and Ducks.* New York: Orange Judd Publishing Co., Inc., 1947.

Kortright, Francis H. *The Ducks, Geese and Swans of North America.* Harrisburg, PA: The Stackpole Co., and Wildlife Management Institute, Washington, D.C., 1976.

Leonard, Dave. *Poultry Feeds and Feeding.* A guide for Peace Corps Volunteers (mimeographed). Arecibo, Puerto Rico: Peace Corps Training Center, January 1968.

Madson, John. *The Mallard.* East Alton, Illinois: Olin Mathieson Chemical Corp., 1960.

May, C.G., ed. *British Poultry Standard.* Poultry World and The Poultry Club of Great Britain, 1971.

National Research Council. *Nutrient Requirements of Poultry,* 7th revised edition. Washington, D.C.: National Academy of Sciences, 1977.

Orr, H.L. *Duck and Goose Raising.* Publication 532, Department of Animal and Poultry Science, Ontario Agricultural College, University of Guelph, Guelph, Ontario, Canada.

Pardee, Roy E., ed. *Raising Ducks for Profit.* Revised edition. The Cory-Keim Publishing Co., 1930.

Penionzhkevich, E.E., ed. *Poultry, Science and Practice.* Vol. I: Biology, Breeds and Breeding. Translated from Russian. Published pursuant to agreement with U.S. Department of Agriculture and the National Science Foundation, Washington, D.C., 1968.

Penionzhkevich, E.E., ed. *Poultry, Science and Practice.* Vol. II: Farming and Production. Trans. from Russ. Publ. pursuant to agreement with U.S. Dept. of Agr. and Nat. Sci. Foundation, Washington, D.C., 1968.

"Predator Problems." *Poultry Press,* Vol. 63, No. 9 (August 1977), p. 10.

"Predator Season's Here." *Poultry Press,* Vol. 63, No. 1 (December 1976), p. 1.

Robinson, John H. *The Growing of Ducks and Geese for Profit and Pleasure.* Dayton, Ohio: Reliable Poultry Journal Publishing Co., 1924.

Salsbury Laboratories. *Salsbury Manual of Poultry Diseases.* Charles City, Iowa: 1971.

Schaible, Philip J. *Poultry: Feeds and Nutrition.* Avi Publishing Co., Inc., 1970.

Sheraw, Darrel. *Successful Duck and Goose Raising.* Pine River, Minnesota: Stromberg Publishing Co., 1975.

Stoddard, H.H. *Domestic Waterfowl.* Hartford, Connecticut, 1885.

Teasley, Mrs. D.O. *All About Indian Runner Ducks.* Middletown, Indiana: Monarch Publishing Co., 1912.

Tegetmeier, W.G. *The Poultry Book.* 1873.

Turk, D.E. and Barnett, B.D. "Cholesterol Content of Market Eggs." *Poultry Science Journal,* Vol. 50, No. 5 (September 1971), p. 1303.

United States Department of Agriculture. *Duck Raising.* Farmer's Bulletin No. 697, 1933.

United States Department of Agriculture. *Raising Ducks.* Farmer's Bulletin No. 2215, 1969.

Valentine, C.S. *The Indian Runner Duck Book,* 1913.

Watt, Bernice K. and Merrill, Annabel L. *Handbook of Nutritional Contents of Foods.* New York: Dover Publications, Inc., 1975. Originally published by the U.S. Department of Agriculture.

Glossary

Bantams: The miniature breeds of ducks.

Biotin: A vitamin found in most feedstuffs (see Appendix B, page 143 for symptoms of a deficiency).

Breed: A subdivision of the duck family whose members possess similar body shape, body size and temperaments and the ability to pass these characteristics on to their offspring.

Breeder ration: Feed used for the production of hatching eggs.

Breeding stock: Adult birds used to produce young.

Broody Hen: A hen that wants to set on eggs. When a hen is broody, her personality undergoes a marked change; she spends much time on the nest, ruffles her feathers when molested, pecks at intruders and quacks incessantly while off the nest.

Concentrated feed: Feeds that are high in protein, carbohydrates, fats, vitamins and minerals, and low in fiber.

Culling: The removal of inferior (crippled, deformed, diseased, low-producing) birds from the flock.

Drake: The male duck.

Duck: In general, any member of the Anatinae family; it is often used specifically in reference to females of the duck family.

Ducklings: Young ducks up until feathers have completely replaced their baby down.

Eclipse molt: A four- to six-week period each year, usually in mid-summer, when the bright plumage of colored adult drakes is replaced with subdued colors similar to those of the hens. This molt occurs for camouflage while the flight feathers are being replaced.

Embryo: The young bird before it emerges from the egg shell.

Esophagus: The tube in which food passes from the mouth to the digestive tract. (Ducks do not have true crops.)

Feed conversion: The ability of birds to convert feed into body growth or eggs. To calculate feed conversion ratios, divide pounds of feed consumed by pounds of body weight or eggs.

Fertility: In reference to eggs, the capability of producing an embryo. Fertility is expressed as a percentage that equals the total number of eggs set minus those that are infertile, divided by the total number set, times 100.

Full-feathered: When a bird has a complete set of feathers.

Gizzard: The muscular organ that grinds the food eaten by birds.

Green ducklings: Young ducks that are managed for fast growth and then slaughtered at seven to eight weeks of age.

Growing ration: Feed that is formulated to stimulate fast growth in ducklings over two weeks old.

Hatchability: The ability of eggs to hatch. Hatchability can be expressed as (1) a percentage of the fertile eggs set (total number of ducklings hatched divided by the number of fertile eggs set, times 100) or (2) a percentage of all eggs set (total number of ducklings hatched divided by the total number of eggs set, times 100).

Hen: The female duck.

Hygrometer: An instrument designed to measure the amount of moisture in the air.

Laying ration: Feed that is formulated to stimulate high egg production.

Lethal gene: A gene that causes the premature death of a developing embryo.

Maintenance ration: Feed used for adult ducks that are not in production.

Molt: The natural replacement of old feathers with new ones.

Niacin: A B-complex vitamin (nicotinic acid). Niacin deficiency causes bowing of the legs.

Nuptial plumage: In colored varieties of ducks, the bright breeding plumage of males exhibited during fall, winter and spring.

Pin feathers: New feathers that are just emerging from the skin.

Pip: The first visible break the duckling makes in the egg shell.

Post-mortem: The thorough examination of a dead bird, usually to determine the cause of death.

Production-bred: Ducks that have been selected for top meat and/or production.

Purebred: Ducks of a specific breed that have not been crossed with other breeds for many generations.

Relative humidity: The ratio of the quantity of water vapor in the air to the greatest amount possible at a given temperature. Therefore, 100 percent relative humidity is total saturation, while 0 percent would indicate the complete absence of moisture.

Roasting chickens: Broilers that are kept beyond eight or nine weeks of age—often up to six months. They attain live weights of 8 pounds or more.

Setting hen: A broody hen that is in the process of incubating a nest of eggs. This term is also sometimes used to identify a hen that has successfully hatched eggs in the past.

Standard-bred: Ducks that have been stringently selected over many generations according to the ideal that is set forth in the Standard of Perfection.

Standard of Perfection: A book containing pictures and descriptions of the physical characteristics desired in the perfect bird of each recognized breed and variety of poultry.

Starting ration: A high-protein feed used the first couple of weeks to get ducklings off to a good start.

Straight run: Young poultry that have not been sexed.

Variety: A subdivision of the breeds. In ducks the varieties within a breed are identified by their plumage color or markings.

Waterfowl: Birds that naturally spend most of their lives on and near water. This term is often used in specific reference to ducks, geese and swans.

Wet-bulb thermometer: A mercury thermometer that has a tubular wick, with one end fitted over the thermometer's bulb and the other end inserted into a container of water. This instrument is used to measure the relative humidity in the incubator.

Index